PROBLEMS:

STATICS AND

MECHANICS OF MATERIALS

ROBERT J. BONENBERGER
JAMES W. DALLY

University of Maryland, College Park

College House Enterprises, LLC
Knoxville, Tennessee

This textbook is intended to provide accurate and authoritative information regarding the various topics covered. It is distributed and sold with the understanding that the publisher is not engaged in providing legal, accounting, engineering or other professional services. If legal advice or other expertise advice is required, the services of a recognized professional should be retained.

The manuscript was prepared using Word 2002 with 11 point Times New Roman font. Publishing and Printing Inc., Knoxville, TN printed this book from camera-ready copy.

College House Enterprises, LLC.
5713 Glen Cove Drive
Knoxville, TN 37919, U. S. A.
Phone   (865) 558 6111
FAX   (865) 558 6111
Email   jdally@collegehousebooks.com
http://www.collegehousebooks.com

ISBN 0-9723567-8-9

CONTENTS

CHAPTER 1 PROBLEMS

1.1 Design a lever that will permit you to lift a weight of 1000 N a distance of 100 mm. In the design, prepare a dimensioned sketch of the lever, state the force that you apply to the lever, and show the analysis proving that you will be able to lift the weight and move it upward by the specified distance.

1.2 The *Sojourner*, a mechanical rover that explored Mars during a mission in 1997, had a mass of 10.5 kg. Determine the weight of *Sojourner* (in newtons) on (a) Earth, (b) Mars. Assume Mars has a mass of 0.64×10^{24} kg and a radius of 3390 km.

1.3 Weather satellites are often placed in geosynchronous orbit around Earth, so that they appear stationary in the sky when viewed from a point on the equator. The orbital altitude for such a satellite is 35,800 km. If a satellite weighs 500 N on the Earth's surface, calculate its weight in orbit. Assume Earth has a mass of 5.976×10^{24} kg and a radius of 6370 km.

1.4 An astronaut weighs 150 lb on the Earth's surface. When orbiting Earth in the space shuttle *Atlantis*, however, the astronaut's weight is measured as 130 lb. Determine the altitude of the shuttle's orbit. Assume Earth has a mass of 5.976×10^{24} kg and a radius of 6370 km.

1.5 Suppose that you are in a space ship traveling in a circular orbit about the Earth, as shown in the figure to the right. Using a spreadsheet, prepare a graph showing your weight as a function of your position relative to the surface of the Earth. Consider radii from $1.0(R_e)$, on the Earth's surface, to $15(R_e)$.

1.6 Determine the gravitational force exerted on the Earth by (a) the sun, (b) the moon. Also find the ratio of the forces (F_{sun}/F_{moon}). Assume masses of 5.976×10^{24} kg for the Earth, 1.990×10^{30} kg for the sun and 7.350×10^{22} kg for the moon. The mean distance (center-to-center) between the sun and the Earth is 149.6×10^6 km and between the moon and the Earth is 384×10^3 km.

1.7 Prove whether or not the blocks, shown in the figure (a) and (b) below, are in equilibrium.

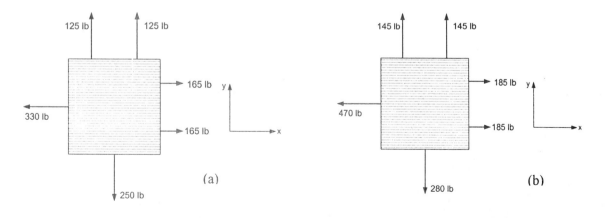

1.8   Prove whether or not the blocks, shown in the figures (a) and (b) below, are in equilibrium.

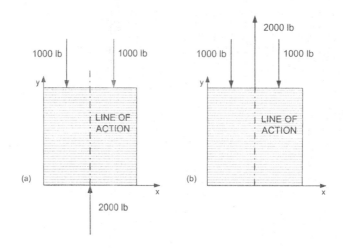

1.9   Draw a free body diagram of an automobile parked on a level street showing the reaction forces on the tires. Also show the active forces on the pavement. Assume that the auto weighs 3,200 lb. State the assumption made in distributing the weight of the auto among the four tires.

1.10   Determine the weight of a body with the following masses in SI units:

(a) 65 kg   (b) 19 kg   (c) 58 kg   (d) 17.3 kg   (e) 34.7 kg   (f) 81.2 kg

1.11   Determine the weight of a body with the following masses in U. S. Customary units:

(a) 7.5 slug   (b) 12 slug   (c) 65 slug   (d) 4.2 slug   (e) 33.7 slug   (f) 49.4 slug

1.12   If a pressure of 3,000 psi acts on a piston with a diameter of 1.75 in., determine the force required to maintain the piston in equilibrium.

1.13   If a pressure of 300 MPa acts on a piston with a diameter of 150 mm, determine the force required to maintain the piston in equilibrium.

1.14   If a force of 3,000 lb acts on a piston with a diameter of 5.0 in., determine the pressure required to maintain the piston in equilibrium.

1.15   If a force of 5,000 N acts on a piston with a diameter of 100 mm, determine the pressure required to maintain the piston in equilibrium.

1.16   If a piston is in equilibrium under a pressure of 1.5 ksi and a force of 10.5 kip, determine the diameter of the piston.

1.17   If a piston is in equilibrium under a pressure of 55 MPa and a force of 675 kN, determine the diameter of the piston.

1.18   List the four basic quantities and give their units in both the SI and U. S. Customary systems.

1.19 Determine the SI equivalent of the masses listed below.

(a) 7.5 slug   (b) 12 slug   (c) 65 slug   (d) 4.2 slug   (e) 33.7 slug   (f) 49.4 slug

1.20 Determine the U. S. Customary equivalent of the masses listed below.

(a) 65 kg (b) 19 kg   (c) 58 kg   (d) 17.3 kg   (e) 34.7 kg   (f) 81.2 kg

1.21 Determine the SI equivalent of the following forces:

(a) 2,400 lb   (b) 425 lb   (c) 688 lb   (d) 18 tons   (e) 52 kip   (f) 95 kip

1.22 Determine the U. S. Customary equivalents for the following forces:

(a) 2,234 N   (b) 89.8 N   (c) 13.2 kN   (d) 96.9 mN   (e) 142.3 μN   (f) 44.8 MN

1.23 Determine the SI equivalent of the following lengths:

(a) 13 ft   (c) 41.3 yards   (e) 6 ft - 2 in.
(b) 153 in.   (d) 16.4 miles   (f) 1.8 miles - 662 yards

1.24 Determine the U. S. Customary equivalents for the following lengths:

(a) 7.4 m   (c) 12.9 cm   (e) 18.8 mm
(b) 500 mm   (d) 13.2 km   (f) 64.3 cm

1.25 Convert the following speeds into SI equivalent units:

(a) 60 mph   (d) 22 ft/s   (g) 16 ft/s
(b) 45 mph   (e) 90 ft/s   (h) 224 ft/s
(c) 30 mph   (f) 580 mph   (i) 187 mph

1.26   Convert the following speeds into U. S. Customary equivalent units:

(a) 120 km/h   (d) 123 m/s   (g) 312 m/h
(b) 61 km/h   (e) 510 m/s   (h) 16.7 m/s
(c) 88 km/h   (f) 17.5 km/s   (i) 43.5 km/s

1.27   Convert the following stresses into SI equivalent units:

(a) 12,000 psi   (d) 132.5 ksi   (g) 194,200 psi
(b) 2,320 psi   (e) 30×10^6 psi   (h) 32 psi
(c) 1,980 psi   (f) 10.5×10^6 ksi   (i) 96,230 psi

1.28   Convert the following stresses into U. S. Customary equivalent units:

(a) 1,000 MPa   (d) 1,400 kPa   (g) 207 GPa
(b) 121 MPa   (e) 144 kPa   (h) 16,430 MPa
(c) 86.4 MPa   (f) 9,642 kPa   (i) 42.3 kPa

1.29 Prepare a short written description of the differences among scalars, vectors, and tensors. Also prepare a sketch illustrating a quantity of each type.

1.30   Consider each quantity listed in Table 1.4 and classify it as a scalar, vector or tensor.

Table 1.4
Unit Conversion Factors

Quantity	U. S. Customary	SI Equivalent
Acceleration	ft/s^2	$0.3048 \ m/s^2$
	in/s^2	$0.0254 \ m/s^2$
Area	ft^2	$0.0929 \ m^2$
	in^2	$645.2 \ mm^2$
Distributed Load	lb/ft	14.59 N/m
	lb/in.	0.1751 N/mm
Energy	ft-lb	1.356J
Force	kip = 1000 lb	4.448 kN
	lb	4.448N
Impulse	lb-s	4.448 N-s
Length	ft	0.3048 m
	in	25.40 mm
	mi	1.609 km
Mass	lb mass	0.4536 kg
	slug	14.59 kg
	ton mass	907.2 kg
Moments or Torque	ft-lb	1.356 N-m
	in-lb	0.1130 N-m
Area Moment of Inertia	in^4	$0.4162 \ x \ 10^6 \ mm^4$
Power	ft-lb/s	1.356 W
	hp	745.7 W
Stress and Pressure	lb/ft^2	47.88 Pa
	lb/in^2 (psi)	6.895 kPa
	ksi = 1000 psi	6.895 MPa
Velocity	ft/s	0.3048 m/s
	in/s	0.0254 m/s
	mi/h (mph)	0.4470 m/s
Volume	ft^3	$0.02832 \ m^3$
	in^3	$16.39 \ cm^3$
	gal	3.785 L
Work	ft-lb	1.356 J

CHAPTER 2 PROBLEMS

2.1 Prepare a free body diagram (FBD) showing the following loads applied to a beam:

(a) concentrated load
(b) uniformly distributed load

(c) non-uniform distributed load
(d) two reaction loads

2.2 Suppose you are to analyze the rope used in a tug of war. Make a section cut of the rope at a location between the two teams and draw a diagram representing the stresses acting on the area exposed by the section cut. Also draw a diagram representing the internal force acting on the area exposed by the section cut.

2.3 Show a graphic representation of a force vector with a magnitude (F) and an orientation (θ) relative to the positive x-axis for:

(a) $F = 7,500$ lb and $\theta = 105°$
(b) $F = 1,400$ N and $\theta = 210°$

(c) $F = 11.2$ kN and $\theta = 48°$
(d) $F = 1.620$ kip and $\theta = 155°$

2.4 Show that the body, illustrated in figures (a) and (b) below, is in equilibrium before and after sliding the 4000 lb force along its line of action.

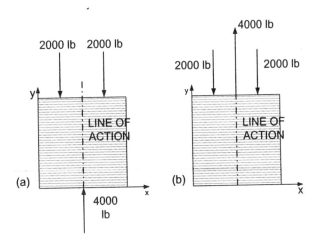

2.5 Draw a disk shaped body and apply a vertical force directed downward at its center. Also show a reactive force at its point of contact with a horizontal support. (a) Describe the condition for equilibrium of the disk. (b) Slide the applied force along its line of action to another location and describe the "new" condition for equilibrium.

2.6 Prepare a drawing showing the reactive forces on the short beam illustrated in the figure to the right.

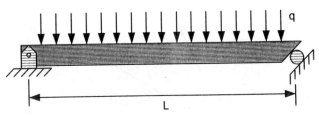

2.7 Prepare three drawings showing the reactive forces on each of the three cylinders in the stack shown in the figure to the right. Note that the cylinders are of the same diameter, and they are bonded together at the three contact points.

2.8 For the vectors **A** and **B**, illustrated in the figure shown to the left, determine the magnitude and direction of the following resultant vectors:

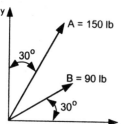

(a) $S_v = A + B$
(b) $D_v = A - B$
(c) $D_v = B - A$.

2.9 For the vectors **A** and **B**, illustrated in the figure shown to the right, determine the magnitude and direction of the following resultant vectors:

(a) $S_v = A + B$
(b) $D_v = A - B$
(c) $D_v = B - A$.

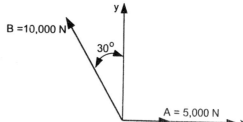

2.10 For the vectors **A** and **B**, illustrated in the figure shown to the left, determine the magnitude and direction of the following resultant vectors:

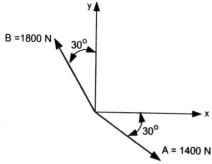

(a) $S_v = A + B$
(b) $D_v = A - B$
(c) $D_v = B - A$.

2.11 Prepare a drawing similar to the one shown below illustrating the vector subtraction $D_v = B - A$.

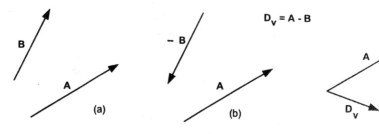

2.12 From the figure shown to the right, determine the Cartesian components of the force vectors: (a) **A** and (b) **B**.

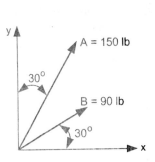

2.13 From the figure shown to the left, determine the Cartesian components of the force vectors:

(a) **A** and (b) **B**.

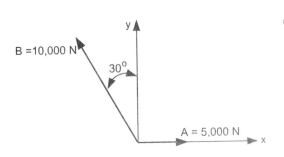

2.14 From the figure shown to the right, determine the Cartesian components of the force vectors:

(a) **A** and (b) **B**.

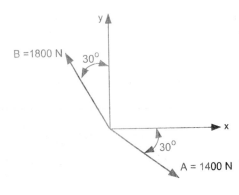

2.15 Using the Cartesian components of vectors **A** and **B,** in the figure shown to the left, determine the magnitude and direction of:

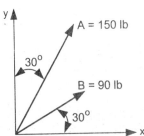

(a) $S_v = A + B$
(b) $D_v = A − B$
(c) $D_v = B − A$.

2.16 Using the Cartesian components of vectors **A** and **B**, in the figure shown to the right, determine the magnitude and direction of:

(a) $S_v = A + B$
(b) $D_v = A - B$
(c) $D_v = B - A$.

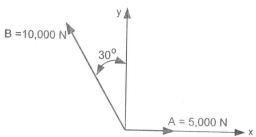

B =10,000 N
30°
A = 5,000 N

2.17 Using the Cartesian components of vectors **A** and **B**, in the figure shown to the left, determine the magnitude and direction of:

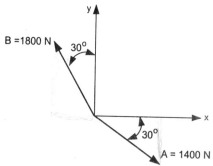

B =1800 N
30°
30°
A = 1400 N

(a) $S_v = A + B$
(b) $D_v = A - B$
(c) $D_v = B - A$.

2.18 Determine the force components along the x and y' axes due to the force of 18,440 lbs that is pulling on the eyebolt shown in the figure to the right.

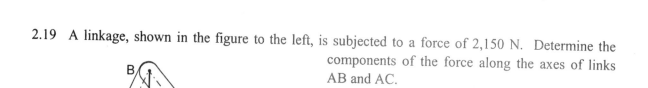

y'
F = 18,440 lb
45°
35°
X

2.19 A linkage, shown in the figure to the left, is subjected to a force of 2,150 N. Determine the components of the force along the axes of links AB and AC.

B
25°
40°
40°
F = 2150 N
A
C

2.20 Determine the angle θ so that the link AB, shown in the figure to the right, has a component of force of 1,250 lb along its axis. The angle φ is fixed at 6.4 degrees.

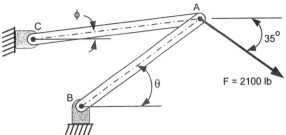

2.21 Determine the vector sum S_v of the three vectors shown in the figure to the right. Specify its direction relative to the x-axis.

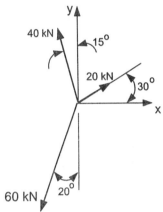

2.22 Determine the magnitude and the direction of the force **F**, shown in the figure to the right if the box is to be lifted from the floor without swinging. The weight of the box is 35 kN.

2.23 If θ is fixed at 55°, in the figure shown to the right, calculate the maximum weight of the box that can be lifted without it swinging when it clears the floor.

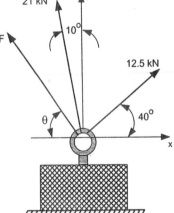

2.24 The gusset plate, illustrated in the figure below, is subjected to four forces due to the attached uniaxial structural members. If the vector sum of the four forces is zero, determine the forces F_1 and F_2.

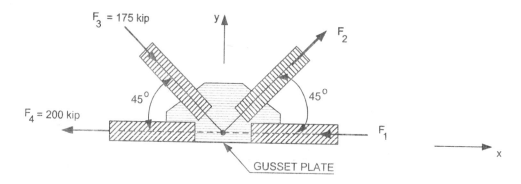

2.25 Three forces act on the bracket, illustrated in the figure to the right. Determine F_1 and θ if the magnitude of one of the Cartesian components of the vector sum $S_{vx} = 16,200$ lb. Note, $S_{vy} = 0$.

2.26 If the vector sum S_v of the three forces is 17,500 lb, and is oriented at 10° relative to the x-axis, determine F_1 and θ. Reference the figure to the right.

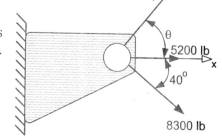

2.27 Define coplanar forces, and show a sketch illustrating four coplanar force vectors.

2.28 Define concurrent forces, and show a sketch illustrating four concurrent force vectors.

2.29 Prepare a sketch of three forces that are:

    (a) coplanar, concurrent   (b) non-coplanar, concurrent   (c) coplanar, non-concurrent

2.30 Suppose you exert a force of 125 N on a single ended lug wrench with a 320-mm long handle to loosen a wheel bolt. What moment (torque) are you applying to the bolt?

2.31 You exert a force of 150 N with each hand on the handles of a double-ended lug wrench to loosen a wheel bolt. If each handle is 225 mm long, what moment (torque) are you applying to the bolt?

2.32 Determine the moment about point O due to the force shown in the figure to the right.

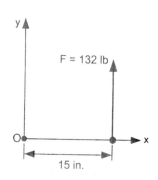

2.33 Determine the moment about point O due to the force shown in the figure to the left.

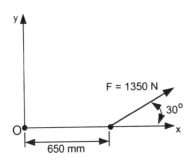

2.34 Determine the total moment about point O due to the two forces shown in the figure to the right.

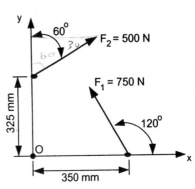

2.35 Express the force **F** or moment **M** in vector format if its Cartesian components are:

(a) $F_x = 2,100$ N, $F_y = -2,300$ N, and $F_z = 1,250$ N
(b) $F_x = 78$ lb, $F_y = -185$ lb, and $F_z = 124$ lb
(c) $M_x = 300$ N-m, $M_y = 400$ N-m, and $M_z = 500$ N-m
(d) $M_x = 740$ in-lb, $M_y = 425$ in-lb, and $M_z = -520$ in-lb

2.36 Determine the magnitude and direction, relative to the positive x-axis, for each of the vectors listed:

(a) $F_x = 1,400$ N, $F_y = 900$ N and $F_z = 0$
(b) $F_x = 355$ lb, $F_y = 520$ lb and $F_z = 0$
(c) $M_x = 1,800$ N-m, $M_y = 1,350$ N-m and $M_z = 0$
(d) $M_x = 655$ ft-lb, $M_y = -335$ ft-lb and $M_z = 0$

2.37 A position vector **r** is constructed in space between the origin (0, 0, 0) and point P. Write **r** as a Cartesian vector and determine the magnitude (r) and coordinate direction angles (α, β, γ) if:

(a) P = (7, 4, 6) in.
(b) P = (110, 65, 70) mm

(c) P = (0.120, 0.155, 0.170) m
(d) P = (−10, 3, −8) ft

2.38 Determine the coordinate direction angles (α, β, γ) for the following forces:

(a) **F** = (4 **i** + 5 **j** − 6 **k**) N
(b) **F** = (−3 **i** + 2 **j** − 6 **k**) lb

(c) **F** = (−3 **i** − 3 **j** + 7 **k**) kN
(d) **F** = (5 **i** − 2 **j** − 8 **k**) kip

2.39 A space force **F** has a known magnitude (F) and two known coordinate direction angles (α, β). Write a vector equation for **F**, specify the three components (F_x, F_y, F_z) and prepare a drawing showing the force vector in a three-dimensional Cartesian coordinate system for:

(a) F = 1,200 N, α = 60° and β = 30°   (c) F = 17 kN, α = 135° and β = 60°
(b) F = 1,500 lb, α = 75° and β = 50°   (d) F = 25 kip, α = 50° and β = 140°

2.40 Write a vector equation for the summation of two space forces F_1 and F_2. The force F_1 is 22.5 kN in magnitude with coordinate direction angles of α_1 = 60°, β_1 = 60° and γ_1 = 135°. The force F_2 is 30.0 kN in magnitude with coordinate direction angles of α_2 = 60°, β_2 = 30° and γ_2 = 90°. Prepare a drawing showing this resultant force vector in a three-dimensional Cartesian coordinate system.

2.41 Determine the angle between the following pairs of space forces:

(a) F_1 = (3 **i** + 6 **j** − 3 **k**) N and F_2 = (−2 **i** + 5 **j** + 7 **k**) N
(b) F_1 = (5 **i** − 1 **j** + 4 **k**) lb and F_2 = (8 **i** − 2 **j** + 3 **k**) lb
(c) F_1 = (−2 **i** + 6 **j** − 3 **k**) kN and F_2 = (−3 **i** + 4 **j** − 4 **k**) kN

2.42 Determine the magnitude of a single force component produced by two space forces (F_1, F_2) that is directed along a line of action which lies in the x–y plane and makes an angle of θ with the positive x-axis:

(a) F_1 = (3 **i** + 6 **j** − 3 **k**) lb; F_2 = (−2 **i** + 3 **j** + 5 **k**.) lb; and θ = 30°
(b) F_1 = (4 **i** + 3 **j** − 2 **k**) kN; F_2 = (3 **i** − 5 **j** + 6 **k**) kN; and θ = 40°
(c) F_1 = (2 **i** − 4 **j** − 2 **k**) kip; F_2 = (−4 **i** + 3 **j** − 5 **k**) kip; and θ = 120°

2.43 A force vector **F**, with components F_x, F_y and F_z is applied to a structure at point Q. Determine the moment of **F** about the origin O of the coordinate system when:

(a) F_x = 4,000 N, F_y = 2,500 N, F_z = 1,500 N and Q = (2.3, 3.4, 4.6) m
(b) F_x = −2,000 lb, F_y = 2,500 lb, F_z = 1,200 lb and Q = (3.4, −2.5, 4.6) ft
(c) F_x = 2.3 kN, F_y = 5.5 kN, F_z = 3.3 kN and Q = (1.4, 3.9, 2.6) m

2.44 If a structure is loaded at point Q with a force **F**, determine the moment M_O about the origin O. Prepare a drawing showing the moment vector M_O in a three-dimensional Cartesian coordinate system. Values for **F** and Q are:

(a) **F** = (125 **i** + 112 **j** + 45 **k**) lb and Q = (2, −3, 15) ft
(b) **F** = (3.8 **i** + 3.3 **j** − 4.5 **k**) kN and Q = (−2.2, 2.6, −1.8) m
(c) **F** = (55 **i** − 89 **j** + 42 **k**) lb and Q = (−3, 4, 8) ft

2.45 The cell phone tower illustrated in the figure to the right is supported by three cables that are maintained with tension forces $F_A = F_B = F_C = 800$ lb. The cables are anchored into the ground plane at locations A, B and C. Because the structural strength and rigidity of the tower is along its axis, we design the cable support system to exhibit a vector sum S_v that is directed along the axis of the tower from point D to the tower support point O. If the tower height $H = 120$ ft and the anchor locations are given in the figure, determine the position for the anchor at point B to achieve the design objective.

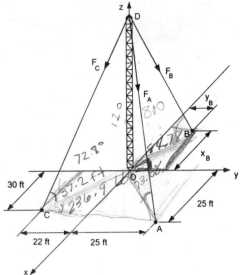

2.46 The cell phone tower, illustrated in the figure to the right, is supported by three cables that are maintained with tension forces $F_A = 800$ lb and $F_C = 1,000$ lb. The cables are anchored into the ground plane at locations A, B and C. Because the structural strength and rigidity of the tower is along its axis, we design the cable support system to exhibit a vector sum S_v that is directed along the axis of the tower from point D to the tower support point O. If the tower height $H = 120$ ft and the anchor locations are given in the figure, determine the force F_B that must be maintained in cable BD to achieve the design objective.

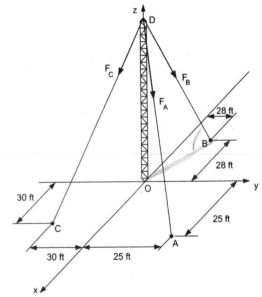

2.47 The boom of a crane extends from its base at point O to its tip at point Q as indicated in the figure below. A three-dimensional coordinate system has been established in this illustration with the point O at the origin and the point Q defined with coordinates $(-1, 3, 8)$. A force with a magnitude of 21 kN is applied by the boom onto a cable that extends from the tip of the boom to point P. Point P is located on the x–y plane with coordinates shown in the figure below. If the coordinates are expressed in meters, determine the following quantities:

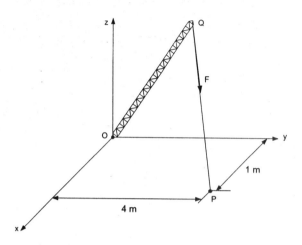

1. The force **F** (in vector form).
2. The force components F_x, F_y, and F_z.
3. The moment $\mathbf{M_O}$ (in vector form).
4. The moment components M_x, M_y, and M_z.
5. The magnitude of the moment (M_O).
6. The unit vector giving the direction of $\mathbf{M_O}$.
7. The angle between **F** and boom OQ.
8. The projection of **F** along boom OQ.

2.48 The boom of a crane extends from its base at point O to its tip at point Q as indicated in the figure below. A three-dimensional coordinate system has been established in this illustration with the point O at the origin and the point Q defined with coordinates $(3, 2, 18)$. A force with a magnitude of 7,500 lb is applied by the boom onto a cable that extends from the tip of the boom to point P. Point P is located on the x–y plane with coordinates shown in the figure below. If the coordinates are expressed in feet, determine the following quantities:

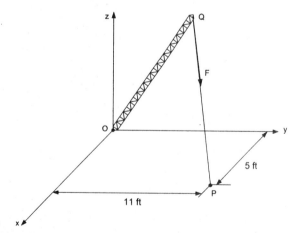

1. The force **F** (in vector form).
2. The force components F_x, F_y, and F_z.
3. The moment $\mathbf{M_O}$ (in vector form).
4. The moment components M_x, M_y, and M_z.
5. The magnitude of the moment (M_O).
6. The unit vector giving the direction of $\mathbf{M_O}$.
7. The angle between **F** and boom OQ.
8. The projection of **F** along boom OQ.

CHAPTER 3 PROBLEMS

3.1 An auto weighing 17,500 N is traveling on a straight highway down a hill with a grade of 3% at a constant velocity of 85 km/h. Construct a FBD of the auto showing all of the forces acting on it. A highway with a grade of 3% elevates 3 m for every 100 m in the horizontal direction.

3.2 A sports utility vehicle weighing 21,500 N is traveling on a straight highway up a hill with a grade of 2% at a constant velocity of 95 km/h. Construct a FBD of the vehicle showing all of the forces acting on it. A highway with a grade of 2% elevates 2 m for every 100 m in the horizontal direction.

3.3 An auto weighing 3,400 lb is traveling on a straight highway up a hill with a 2% grade at a constant velocity of 65 MPH. Construct a FBD of the auto showing all of the forces acting on it. A highway with a grade of 2% elevates 2 ft for every 100 ft in the horizontal direction.

3.4 A sports utility vehicle weighing 4,500 lb is traveling on a straight highway down a hill with a 3.5% grade at a constant velocity of 70 MPH. Construct a FBD of the auto showing all of the forces acting on it.

3.5 Prepare a sketch illustrating the force system described below and write the relevant equilibrium relations for it.

   (a) Coplanar and concurrent   (c) Coplanar and non-concurrent
   (b) Non-coplanar and concurrent   (d) Non-coplanar and non-concurrent

3.6 Construct a FBD for the mass and two-cable assembly shown in the figure to the right. Also determine the tension forces in cables AC and BC.

3.7 Construct a FBD for the mass and two-cable assembly shown in the figure to the left. Also determine the tension forces in cables AC and BC.

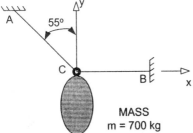

3.8   Construct a FBD for the mass and three-cable assembly shown in the figure to the right. Determine the tension forces in cables AC, BC and CD. Assume cable CD supports 30% of the mass.

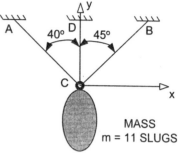

3.9   Determine the forces in cables AC and BC that support the signal lights at a traffic intersection as illustrated in the figure to the left. The signal lights have a mass m = 40 kg. The dimensions for the cable arrangement are given in meters.

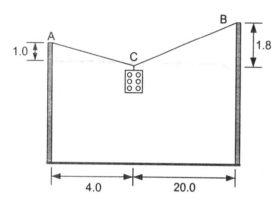

3.10   A container A that holds a 1.0 ton weight ton is lifted from the floor with a cable and pulley system shown in the figure to the right.   Determine the equation for the force F required to lift the container as a function of α. Evaluate this equation to determine the force F if $\alpha = 20°$. Comment on the effectiveness of this cable pulley system.

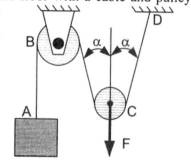

3.11   A container holding a mass m = 850 kg is lifted from the floor with a cable and pulley system shown in the figure to the right. Determine the force F required to lift the container as a function of α. Evaluate this equation to determine the force F if $\alpha = 17°$. Comment on the effectiveness of this cable pulley system.

3.12   Using the pulley and cable arrangement shown in the figure to the right, design a lifting arrangement that requires only a force of 1,000 lb to lift a container weighing 1,800 lb. Assume that the pulleys are frictionless.

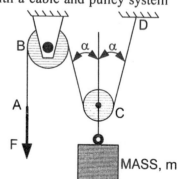

3.13 Using pulleys and cables, design a lifting arrangement that requires only a force of 500 lb to lift a one-ton container. Assume that the pulleys are frictionless.

3.14 For the pulleys and cable arrangement shown in Problem 3.11, determine the force required to lift a 4.5 kN weight as a function of the angle α the cables make with pulley C. Let the angle α vary from zero to 90°. It is suggested that you use a spreadsheet for these calculations and prepare a graph showing the results.

3.15 A differential hoist is illustrated in the figure to the right. At the top, two pulleys with radii r_1 and r_2 are fastened together so they turn as a single unit. A continuous cable passes around the smaller pulley, with a radius r_2, and then around the lower pulley with a radius r_3, and finally around the larger pulley with a radius r_1. We assume that the pulleys are frictionless relative to the shafts passing through their hubs. We also assume that the cables do not slip in the grooves of the pulleys. If the diameter of the lower pulley is $D = 2r_3 = r_1 + r_2$, determine the equation for the applied force F in terms of the load W, which is maintained in equilibrium.

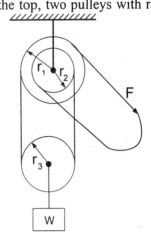

3.16 An artist has designed a bell using a long piece of pipe as the vibrating sound source. The pipe is to be supported using the wire ring arrangement shown in the figure to the left. Determine the force in each of the wires. The pipe has a mass of 43 kg.

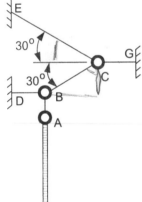

3.17 The pulley and cable, shown in the figure to the right, support a mass m = 12 slug suspended at point A. The cable B-A-C is 65 ft long and sags prior to the addition of the pulley and the container.
   (a) If the pulley diameter is small relative to the span of 50 ft between the walls at B and C, determine the height h associated with the equilibrium position of the pulley and container.
   (b) If the pulley and container are released from point C, describe the motion, which you would observe as the system reaches equilibrium.
   (c) If the pulley and container are released from point B, describe the motion, which you would observe as the system reaches equilibrium.

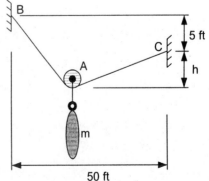

3.18 Two masses m_B and m_C are supported by the cable arrangement shown in the figure to the right. Determine the mass m_C and the forces in the cables AB, BC and CD if the mass $m_B = 16.0$ kg.

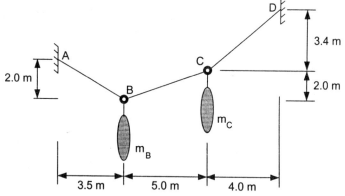

3.19 A bungee cord is stretched across a river as shown in the figure below. A stuntman weighing 140 lb is to cross the river by tightrope walking on the bungee cord. If the cord has a linear spring constant k = 60 lb/in., determine if the stuntman will remain dry.

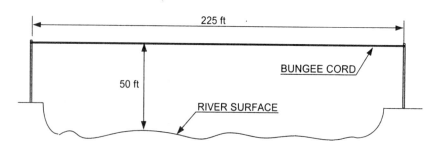

3.20 For the conditions described in Problem 3.19, determine the spring constant of the bungee cord if the stuntman is to barely wet the soles of his shoes.

3.21 Three springs are connected in series as shown in the figure to the right. Determine the extension of the entire arrangement if a force of 23.4 lb is applied to the last spring. Also determine the effective spring rate for the three springs in this series arrangement.

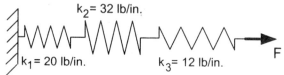

3.22 Three springs are connected in series as shown in the figure to the left. Determine the extension of the entire arrangement if a force of 70 N is applied to the last spring. Also determine the effective spring rate for the three springs in the series arrangement.

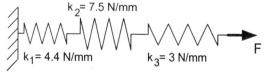

3.23 For the parallel spring arrangement determine the deflection of the lower bar shown in the figure to the right. The spring rates $k_A = 12$ N/mm. and $k_B = 23$ N/mm. The force F = 1145 N. Also determine the effective spring rate for the three springs in parallel.

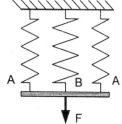

3.24 Three springs are used to support the rigid bar with a mass m as illustrated in the figure to the right. If $k_1 = 20$ lb/in., $k_2 = 28$ lb/in., $k_3 = 36$ lb/in. and m = 1.3 slug, determine the amount each spring elongates. The bar is constrained and remains parallel to the top surface to which the springs are attached.

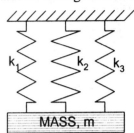

3.25 A cable and a spring are connected together at point C as shown in the figure to the left. Determine the equilibrium position of point C if a force F = 400 N is applied. The undeformed length L of both the cable and the spring before the application of the load is shown in the figure together with the spring rate k.

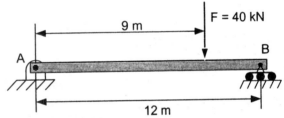

3.26 For the beam, shown in the figure to the right, solve for the reactions at the pin and clevis and the roller.

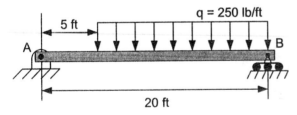

3.27 For the beam, shown in the figure to the left, solve for the reactions at the pin and clevis and the roller.

3.28 For the beam, shown in the figure to the right, solve for the reactions at the pin and clevis and the roller.

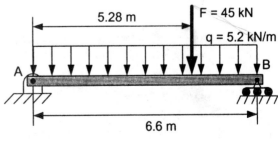

3.29 For the beam, shown in the figure to the right, solve for the reactions at the built-in end of the cantilever beam.

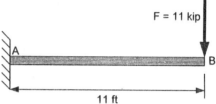

3.30 For the beam, shown in the figure to the left, solve for the reactions at the built-in end of the cantilever beam.

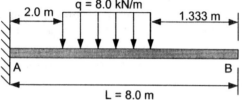

3.31 For the beam, shown in the figure below, solve for the reactions at the two supports.

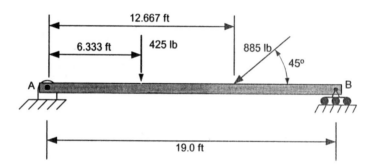

3.32 For the beam, shown in the figure below, determine the reactions at the two supports. The dimensions of the beam and the mass are given in the table below. Note that the arm CD is rigidly attached to the beam. Also, the radii of the two pulleys are the same and equal to 0.25 m.

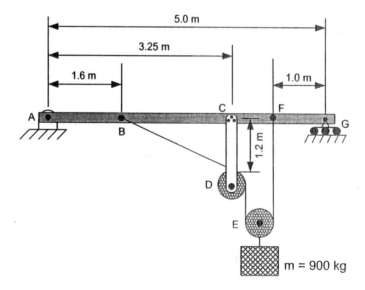

3.33 For the truss, shown in the figure to the right, determine the reactions at the two supports.

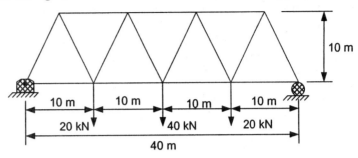

3.34 A long slender pole is to be erected by a winch and cable arrangement that is mounted on a

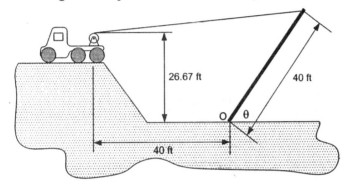

truck, as illustrated in the figure to the left. If the pole weighs 2,000 lb and is 40 ft long, determine the reactions at point O and the force in the cable. Consider angles θ varying from 20° to 75°. A pivot is attached to the base of the pole so that it may be rotated without a resisting moment (i.e. acts like a pin).

3.35 For the crane, described in the figure to the right, determine the maximum load W_L that can be lifted before the crane tips.   Note, the crane weights 30 kN.

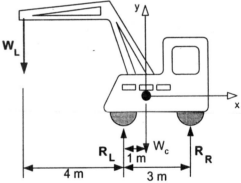

3.36 For the crane, illustrated in the figure to the left, determine the maximum load that can be lifted before the crane tips.   Note, the crane weights 3.5 ton.

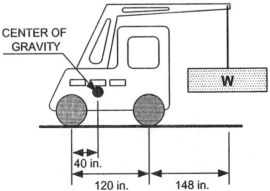

3.37 Determine the reaction forces and moments at the support for the stadium structure shown in the figure to the right.

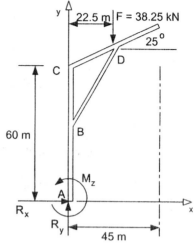

3.38 Determine the reaction forces at pins A and B that support the gusset plate shown in the figure to the left. Assume that the vertical components of the reaction forces at the pins are equal (i.e. $R_{Ay} = R_{By}$)

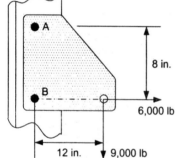

3.39 For the tension bar, shown in the figure to the right, determine the internal tension force P at sections A-A and B-B.

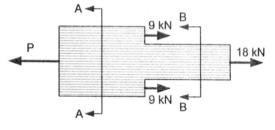

3.40 For the beam, shown in the figure to the right, determine the internal shear force V and the internal moment M at the position $x = L/2$.

3.41 For the beam, shown in the figure to the right, determine the internal shear force V and the internal moment M as a function of position for $0 < x < L$. Prepare a graph of M and V as a function of x.

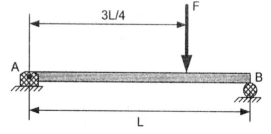

3.42 For the beam, shown in the figure to the right, determine the internal shear force V and the internal moment M at position x = L/3. Present your results in terms of q and L, which are known quantities.

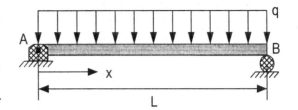

3.43 For the beam, shown, in the figure to the right, determine the internal shear force V and the internal moment M as a function of position x. Prepare a graph of M and V as a function of x.

3.44 For the beam, shown in the figure to the right, determine the internal shear force V and the internal moment M at position x = 3L/4. Present your results in terms of F and L, which are known quantities.

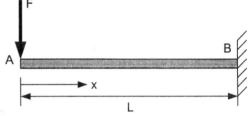

3.45 For the beam, shown in the figure to the right, determine the internal shear force V and the internal moment M as a function of position x. Prepare a graph of M and V as a function of x.

3.46 For the beam, shown in the figure to the lower right, determine the internal shear force V and the internal moment M at the position x = 12 ft.

3.47 For the beam, shown in the figure to the right, determine the internal shear force V and the internal moment M as a function of position x. Prepare a graph of M and V as a function of x.

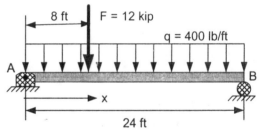

3.48 For the C-clamp, shown in the figure below, determine the internal forces P and V and the internal moment M_z at section A-A. The force F applied by the screw is 250 lb and the dimension D, which defines the distance from the screw to the centerline of the C-clamp is 1.5 in.

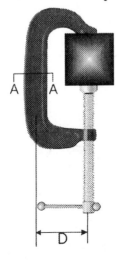

3.49 A new factory for heavy machinery has an assembly line with component parts stored above floor level. The components used in assembly are in elevated bins (#1, #2, and #3) as shown in the figure to the right. The elevated bins are suspended from a frame ABCD. Prepare a FBD of the left portion of the frame from Section A-A to the roller support at point A. Determine the internal forces and moment at section A-A as a function of the loads W_1, W_2 and W_3. Locate Section A-A at position $x = 3L/8$ and consider the loads W_1, W_2 and W_3 as known quantities. Assume that the centers of each bin are located at the following points: $x_1 = L/4$, $x_2 = L/2$ and $x_3 = 3L/4$.

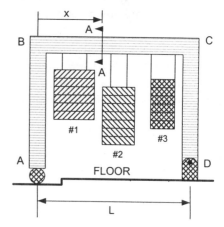

CHAPTER 4 PROBLEMS

4.1 A steel music wire (Gage No. 10 with a diameter of 0.024 in.) is employed to support an axial load P of 8.64 lb. If the wire was initially 22 ft in length, determine its final length and the stress and strain in the wire.

4.2 A steel music wire is tightened on a guitar by rotating the wire supporting post through an angle of 11°. If the post has a diameter d = 0.25 in. and the length of the wire is 32 in., determine the strain induced in the guitar string.

4.3 Referring to Problem 4.2, determine the additional stress induced in the guitar string as it is tightened.

4.4 A hemp rope 15 m long and 45 mm in diameter is supported from one of the roof beams in a gymnasium. Three gymnasts climb this rope together. The lead gymnast has a mass of 55 kg, the next 72-kg and the trailing gymnast 52 kg.

    (a) Determine the maximum stress in this rope.
    (b) Determine the stress in the rope at a location between the lead and intermediate gymnast.
    (c) Determine the stress in the rope at a location between the intermediate and trailing gymnast.

4.5 A suspension bridge is to carry a roadway and traffic that may weigh up to 5,000 ton. If the safe load that can be imposed on the wire rope to be used in the construction is 175 ton, specify the number of cables to be employed. Justify your answer.

4.6 Describe the constraints on the relation $\sigma = E\varepsilon$.

4.7 Determine the design load that can be specified for Gage No. 00 steel wire with a strength of 530 MPa, if the design specification calls for the safety factor of 2.5.

4.8 Discuss the factors to consider in establishing the safety factor in a design specification.

4.9 For the cable-pulley-mass arrangement shown in the figure to the right, determine the stresses in the cable. The effective cable diameter d is 0.500 in.

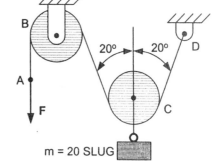

4.10 Cold drawn steel alloy wire exhibits an ultimate tensile strength of 600 MPa and a yield strength of 500 MPa. If this alloy is employed, specify the gage of the wire if it is to support a design load of 7.5 kN with a safety factor of 3.2. Specify the gage number required based on both yielding and rupturing as modes of failure.

4.11 Determine the safety factor for the wire CD in the cable-weight arrangement shown in the figure
to the right. The steel cables are each 10 m long, exhibit a yield
strength of 500 MPa and have an effective cross sectional area of
8 mm². Assume that the forces in the horizontal wires are equal
in magnitude and opposite in direction ($F_{CB} = -F_{CA}$). Also
assume $F_{CB} = (0.18)F_{CD}$.

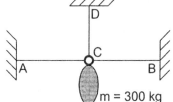

4.12 Determine the mass m, shown in the diagram to the right, that is
required to yield the cable CD if the cables are each 22 ft long,
exhibit a yield strength of 80ksi and have an effective cross sectional area of 0.015 in.² Assume
that the forces in the horizontal wires are equal in magnitude and opposite in direction ($F_{CB} = -$
F_{CA}). Also assume $F_{CB} = (0.25)F_{CD}$. Is the cable-weight arrangement stable after yield? Why?

4.13 The wire-mass system, shown in the figure to the left, is constructed using steel wire. Specify a
suitable alloy and the gage number for the wires if the criterion
for failure is based on yielding. Use the same diameter for both
wires. The safety factor imposed on the design is 2.9.

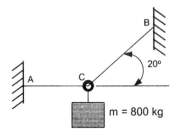

4.14 The wire-mass system, shown in the figure to the right, is constructed using aluminum wire.
Specify a suitable alloy and the gage number for the wires if the
criterion for failure is based on ultimate tensile strength. Use the
same diameter for both wires. The safety factor imposed on the
design is 1.8.

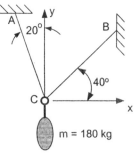

4.15 The wire-mass system ,shown in the figure to the left, is constructed using steel wire fabricated
from 1010 A. A safety factor of 2.9, based on the ultimate tensile
strength, is to be employed for both wires. Find the maximum
weight that can be lifted (in pounds) and specify the required
gage numbers for each wire.

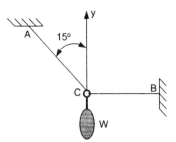

4.16 Describe in an engineering brief the differences between a wire rope and a rod.

4.17 Describe in an engineering brief the similarities between a wire rope and a rod.

4.18 A long thin steel bar with a length of 1.8 m, a width of 50 mm, and a thickness of 20 mm is subjected to an axial force of 22 kN. Determine the tensile stress and the axial deformation of the bar.

4.19 A bar fabricated from steel with a tensile strength of 54.0 ksi is subjected to an axial tensile force of 20 kip. The bar is designed with a safety factor of 3.2. Determine the design stress for the bar and the required cross sectional area.

4.20 A bar fabricated from an aluminum alloy with a tensile strength of 400 MPa is subjected to an axial tensile force of 75 kN. The bar is designed with a safety factor of 2.5. Determine the design stress for the bar and the required cross sectional area.

4.21 A bar fabricated from brass with a tensile strength of 247 MPa is subjected to an axial compressive force of 76 kN. The bar is designed with a safety factor of 2.8. Determine the design stress for the bar and the required cross sectional area. Assume the tensile and compressive stresses are equal.

4.22 A key used to lock a gear onto a shaft is subjected to a shear force of 8.0 kip. If the key is 0.25 in. wide by 0.375 in. high and 1.5 in. long, determine the shear stress acting on the key. Draw a free body diagram showing this shear stress.

4.23 If the key in Problem 4.22 is machined from a steel alloy with a tensile yield strength of 48,000 psi, determine the safety factor for the key.

4.24 A rectangular bar, shown in the figure to the right, is subjected to an axial tensile force F = 30 kN. Determine the normal stress and the shear stress on an inclined plane if the angle of the section cut is θ = 30°.

4.25 A rectangular bar, similar to the one shown in the figure to the right, is adhesively bonded along a 20° section cut. The normal stress in the bond line is limited to 2.5 ksi and the shear stress to 1.5 ksi. If the bar has a cross sectional area A = 1.75 in.^{2} and a safety factor of 3.0 is specified, determine the largest axial force that can be applied to the bar.

4.26 A rod with a circular cross section with a diameter d is fabricated from two pieces as illustrated in the figure to the right. If a force F is applied to the bar, derive the expression for the normal and shear stress acting in the plane of the adhesive joint as a function of angle ϕ.

4.27 For the bar illustrated in the figure to the right, determine the shear stresses in the adhesive joint if the angle ϕ is varied from 0° to 90°. The axial force applied to the bar is 2.0 kN and its diameter is 20 mm. Also compute the normal stresses acting on the adhesive joint. Hint: Use a spreadsheet to determine the shear and tensile stresses in the adhesive joint for the range of ϕ specified.

4.28 A tensile bar, defined in the figure to the right, has a cross sectional area of 175 mm^2 and is subjected to a tensile force F. The stresses on the inclined plane A—B are σ_θ = 81 MPa and τ_θ = −27 MPa. Determine the stress σ_x, the angle θ and the force F.

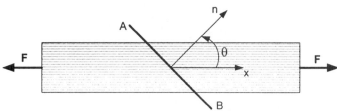

4.29 The tensile bar, defined in the figure to the right, has a cross sectional area of 175 mm^2 and is subjected to a tensile force F. The stresses on the inclined plane A—B are σ_θ = 81 MPa and τ_θ = −27 MPa. Determine the shear and normal stresses acting on an inclined plane with $\theta = 40°$.

4.30 For the tensile bar in Problem 4.29, prepare a graph that shows the stresses σ_θ and τ_θ as a function of the angle of the inclined cut as θ varies from 0 to 90°.

4.31 Consider the tapered bar presented in the figure below. The tapered bar is subjected to an axial force of 500 kN. Prepare a graph of the axial stress σ_x as a function of x as it varies from zero to 4 m.

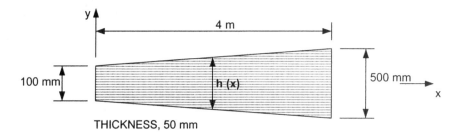

THICKNESS, 50 mm

4.32 Consider the tapered bar presented in Problem 4.31. If the bar is fabricated from an aluminum alloy, determine the extension of the bar when it is subjected to an axial tensile force of 500 kN.

4.33 Consider the tapered bar presented in the figure below. If the bar is fabricated from an aluminum alloy, determine the extension of the bar when it is subjected to a tensile force of 120 kN. In this solution, do not employ Eq. (4.19). Instead, use Eq. (4.18) and perform a numerical integration on a spreadsheet.

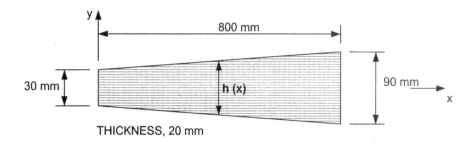

THICKNESS, 20 mm

4.34 Describe the procedure employed to solve for the stresses and deflection in a stepped bar subjected to axial loading.

4.35 For the stepped bar, illustrated in the figure to the right, determine the stresses in both portions of the bar and its total deflection. The bar is fabricated from steel.

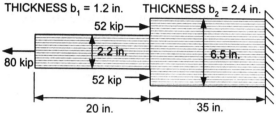

4.36 A tensile bar 40 in. long, shown in the figure below, is subjected to an axial force of 10 kip. Determine: (a) the nominal stress, (b) the stress concentration factor K, and (c) the maximum stress.

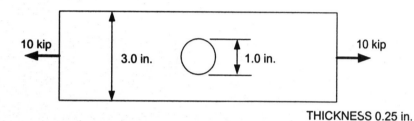

THICKNESS 0.25 in.

4.37 A stepped tensile bar, illustrated in the figure below, with fillets at the transition between the small and the large section is subjected to an axial tensile force of 75 kN. Determine the maximum stress at the fillets.

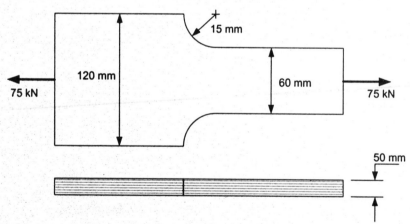

4.38 A scale model of a large structure has been fabricated from steel and tested. A strain gage on one member of the structure indicated an axial strain of $\varepsilon = 1,450$ μm/m. Determine the stress in the corresponding member of the prototype. The numerical parameters defining the scaling factors for the loads and the size of the structural member are $w_m = 1.6$ mm, $w_p = 80$ mm, $b_m = 2.0$ mm, $b_p = 100$ mm, $L_m = 14$ mm and $L_p = 7.0$ m. The scaling factor $L = 1/10,000$.

4.39 Suppose that a model of a structure is fabricated from members formed from sheet aluminum. The prototype structure is to be fabricated from steel with an open span of 200 feet. The model is geometrically scaled so that its span is four feet. The capacity of the live load on the prototype is 150 lb/ft, and the model is loaded with 2.5 lb/ft. If the model deflects a distance of 0.220 in. under full load at the center of the span, determine the deflection of the prototype under the design load.

4.40 The tensile bar, presented in the figure to the right, is fabricated with an adhesive joint inclined
at an angle ϕ to the axis of the bar.
Determine the optimum angle ϕ for
the inclined plane if the stresses in
the adhesive are not to exceed either
3.5 ksi in tension or 2.8 ksi in shear.
Also determine the maximum force
that can be applied to the member without exceeding the stresses in the adhesive joint.

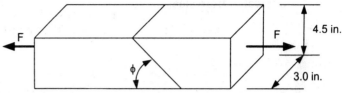

4.41 A structural member is fabricated from a solid round bar of steel with a diameter d = 50 mm. If
the member is 6.0 m long, determine the maximum axial force that can be applied if the axial
stress is not to exceed 175 MPa and the total elongation is not to exceed 0.14% of its length.

4.42 A steel tie rod with a diameter of 1.0 in. and length of 34.0 in. is employed to compress a brass
bushing with an outside diameter of 3.0 in. and a length of
24.0 in. as shown in the figure to the right. Determine the
minimum wall thickness of the bushing if the deflection of
the tie rod is limited to $\delta = 0.020$ in. Note that F = 10 kip.

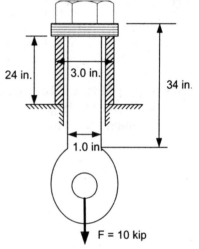

4.43 A short column with a height of 2.0 m is fabricated by adhesively bonding aluminum faceplates
to a core of plastic foam as shown in the figure to the
left. The foam plastic core has a square cross section
with an elastic modulus of 1200 psi. Determine the
stresses in the aluminum plates and the plastic foam.
Also, determine the displacement of the column
under the action of the applied force F = 35 kN.

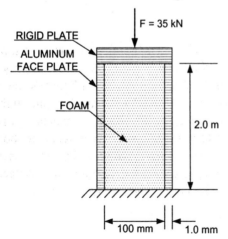

4.44   A grade 8 steel bolt, 1.0 in. in diameter (14 threads/in.) is employed to clamp a brass bushing between two rigid platens as illustrated in the figure below. The brass bushing has a 2.5 in. outside diameter and 1.5 in. inside diameter and is 12.00 in. long. After the unit is assembled with a snug fit, the nut is tightened by 1/3 of a turn. Determine the axial stresses in the bolt and the bushing. Also determine the deflection of the brass bushing.

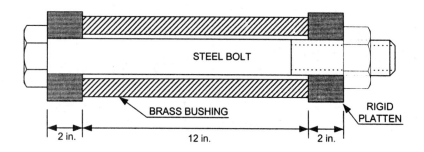

4.45   Twenty-five steel reinforcing rods with a ¼ in. diameter are placed within a high strength concrete column with a square cross section that supports an applied force of 50 kip. Determine the stresses in the steel reinforcing bars and the concrete. Also determine the amount of deflection of the column. Reference the figure to the right.

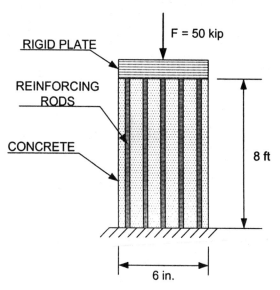

4.46   A long aluminum rod is connected to a shorter brass cylinder with a bolted flange as shown in

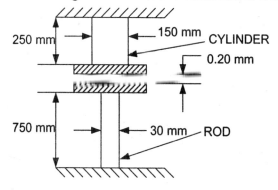

the figure to the left. Prior to assembly a gap of δ = 0.20 mm occurred between the flange plates. Bolts were inserted and tightened bringing the flange plates together. Determine the stresses induced in both the rod and the cylinder by the assembly operation. Also determine the displacement of the face of each flange.

4.47 A copper and stainless steel rod are assembled between two rigid walls at an ambient temperature of 20° C as shown in the figure to the right. If the temperature is increased by an amount ΔT = 120 °C, determine the thermal stresses induced in each rod. Also determine the change in length of each member.

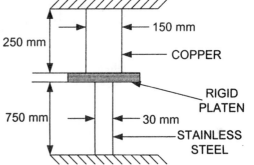

4.48 A high strength steel bolt with a diameter of 1.0 in. passes through an aluminum bushing with a cross sectional area of 3.0 in.² as shown in the figure below. The unit is assembled with a snug fit at an ambient temperature 70 °F. If the temperature of the assembly increases by ΔT = 60 °F, determine the thermal stresses induced in the bolt and the bushing.

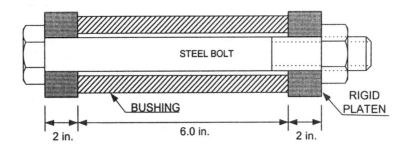

4.49 A three bar suspension system, shown in the figure to the right, is assembled at an ambient temperature of 20° C. Determine the axial stresses in each bar after a force F of 100 kN is applied to the rigid platen and the temperature is increased by ΔT = 80 °C. The bars labeled A are fabricated from aluminum with a cross sectional area of 750 mm² and the bar labeled B is fabricated from stainless steel with a cross sectional area of 1,250 mm².

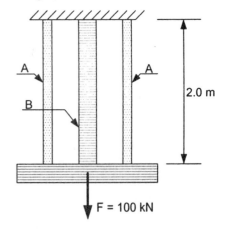

CHAPTER 5 PROBLEMS

5.1 Sketch a stress-strain diagram for a relatively brittle material and identify the ultimate tensile strength on the diagram.

5.2 Sketch a stress-strain diagram for a relatively ductile material and identify the yield strength and ultimate tensile strength on the diagram.

5.3 Sketch a stress-strain diagram for a ductile material that does not exhibit a well defined yield point. Identify on the diagram the yield strength and the ultimate tensile strength.

5.4 Write a test plan for conducting a standard tensile test.

5.5 Determine the yield strength and the ultimate tensile strength for the material represented by the stress-strain diagram shown in the figure to the right.

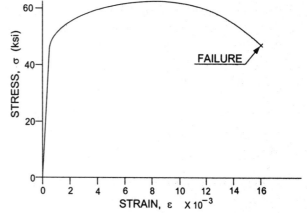

5.6 Determine the yield strength and the ultimate tensile strength for the material represented by the stress-strain diagram shown in the figure to the left.

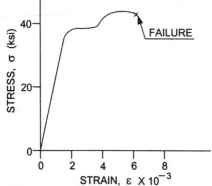

5.7 Determine the percent elongation and the percent reduction in area if measurements of $L_f = 66.4$ mm and $d_f = 10.52$ mm were made during a standard tensile test. The initial length and diameter of the standard tensile specimen is $L_o = 50$ mm and $d_o = 12.7$ mm.

5.8   In conducting a tensile test of an aluminum alloy, you adjust the signals from the load cell and the extensometer to zero and then load the specimen. You dwell at several loads in the elastic region of the stress-strain response, and record readings of the load F and the stretch δ. The values of F and δ are plotted on an F-δ diagram to obtain the slope $\Delta F/\Delta\delta = (16{,}000 \text{ N})/(100 \times 10^{-3} \text{ mm})$. Determine the elastic modulus of the aluminum alloy if the initial length and diameter of the tensile specimen is $L_o = 50$ mm and $d_o = 12.7$ mm.

5.9   If a steel alloy exhibits a Poisson's ratio of $\nu = 0.30$, determine the diameter of a tensile specimen when it is subjected to an elastic stress of $\sigma = 52.6$ ksi. The dimensions L_o and d_o for the specimen are 2.00 in. and 0.500 in., respectively.

5.10   A long (3 m) sheet of an aluminum alloy is stretched in a manufacturing process until it is 3.8 m in length. If the sheet was initially 0.6 m wide and 2 mm thick, determine its new width and thickness. In the plastic regime, Poisson's ratio is 0.5 for all metallic materials because volume is conserved in the plastic deformation process.

5.11   A long (3.8 m) sheet of an aluminum alloy is stretched in a manufacturing process until it is 4.6 m in length. If a central hole initially 75 mm in diameter was drilled in the sheet prior to stretching, determine its new dimensions. In the plastic regime, Poisson's ratio is 0.5 for all metallic materials because volume is conserved in the plastic deformation process.

5.12   Prove that volume of an object is conserved in a uniaxial stress state if $\nu = \frac{1}{2}$.

5.13   If a steel alloy is linearly elastic until the applied axial stress equals the yield strength, determine the strain at yield for a structural steel with a yield strength of 38 ksi and a modulus of elasticity of 30×10^6 psi.

5.14   A long, thin, aluminum-alloy bar is subjected to an axial stress of 227 MPa. Determine the strain in the axial and transverse directions of the bar.

5.15   A spherical pressure vessel, fabricated from a steel alloy, is subjected to a biaxial stress field with $\sigma_x = \sigma_y = 20.6$ ksi. Determine the strains ε_x and ε_y.

5.16   An electrical strain gage is placed with an arbitrary orientation on a spherical pressure vessel fabricated from a titanium alloy. The gage provides a strain measurement $\varepsilon = 1{,}140 \times 10^{-6}$. Determine the stresses σ_x and σ_y in terms of MPa.

5.17   Prove that the ratio of hoop strain to axial strain for a cylindrical pressure vessel is given by:

$$\varepsilon_h/\varepsilon_a = (2 - \nu)/(1 - 2\nu).$$

5.18   An orthogonal pair of electrical strain gages is placed on a cylindrical pressure vessel fabricated from an aluminum alloy. The strain gages provide measurements of the axial strain $\varepsilon_a = 264 \times 10^{-6}$ and hoop strain $\varepsilon_h = 1{,}232 \times 10^{-6}$. Determine the axial stress σ_a and the hoop stress σ_h in terms of MPa.

5.19   A standard tensile specimen ruptured with an applied force $F = 11.3$ kip. Your measurement of the diameter at the neck region after failure gives $d_f = 0.366$ in. Determine the engineering stress and the true stress at failure.

5.20 Determine the engineering strain and the true strain for a tensile specimen with $L_o = 2.00$ in. if the final length, L_f is given in the table shown below:

Final Length L_f (in.)	Engineering Strain	True Strain
2.1		
2.2		
2.3		
2.4		
2.5		
2.6		
2.7		
2.8		
2.9		

5.21 Determine the engineering strain and the true strain for a tensile specimen with $L_o = 50$ mm if the final length L_f is given in the table below:

Final Length L_f (mm)	Engineering Strain	True Strain
53		
56		
59		
62		
65		
68		
71		
74		
77		

5.22 Complete the conversion table that relates engineering strain and true strain.

Engineering Strain	True Strain
0.001	
0.002	
0.005	
0.010	
0.020	
0.050	
0.100	
0.200	
0.500	
1.000	

5.23 The maximum stress imposed at the root of a tooth on a spur gear is $\sigma_{max} = 78.9$ ksi and the minimum stress $\sigma_{min} = 0$. Neglecting the influence of the cyclic mean stress on the fatigue life, determine the anticipated life of the gear, if the $S_f - N$ curve presented to the right characterizes its fatigue properties.

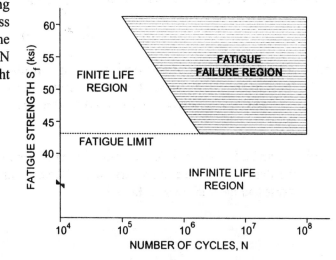

5.24 A rule for approximating the fatigue strength is $S_f = S_u/2$. Using this rule, determine the maximum stress that can be imposed on a structural member with a dynamic load that is twice the static load as illustrated in the figure below. The structural member is fabricated from steel with $S_u = 340$ MPa. Neglect the influence of the cyclic mean stresses on the fatigue behavior.

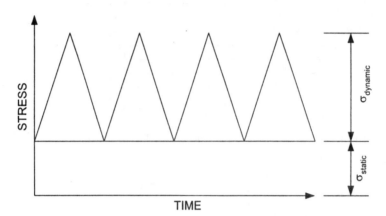

5.25 Reconsider Problem 5.23 taking into account the effect of the cyclic mean stress.

5.26 Reconsider Problem 5.24 taking into account the effect of the cyclic mean stress.

5.27 Repeat Problem 5.24 with the dynamic load equal to the static load.

5.28 Repeat Problem 5.24 with the dynamic load equal to one half of the static load.

5.29 Beginning with Eq. (5.9) derive Eq. (5.10).

5.30 Convert the stress-strain curve shown in the figure to the right to a true stress-true strain curve.

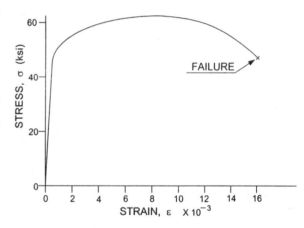

5.31 Convert the stress-strain curve shown in the figure to the left to a true stress-true strain curve.

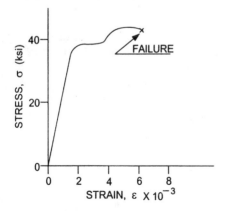

5.32 At the proportional limit in a tensile test, a 2 in. gage length had elongated by 0.003 in. and the diameter of the standard specimen was smaller by 0.00026 in. The load measured by a load cell fitted to the tensile machine at the proportional limit was 5,000 lb. Determine the modulus of elasticity, Poisson's ratio and the proportional limit for this material.

5.33 A rod with a diameter of 1.00 in. and length of 8.0 ft undergoes an extension of 0.220 in. when subjected to an axial force of 54.0 kip. The diameter of the rod decreases by 0.0007 in. at this load. Determine the modulus of elasticity, Poisson's ratio and the shear modulus for the rod's material.

5.34 A standard tensile specimen was fitted with an extensometer with a 2.00-inch gage length and tested until failure. The force and extension measured during the test is presented in the table below. The diameter of the specimen at the fracture neck was 0.413 in. Analyze the data using a spreadsheet and determine the following quantities.

    a.   Modulus of elasticity
    b.   Yield strength (0.2% offset)
    c.   Ultimate strength
    d.   Fracture stress
    e.   Strain at fracture
    f.   Percent elongation
    g.   Percent reduction in area.

Load (kip)	Extension (in.)	Load (kip)	Extension (in.)
0	0	10.8	0.251
2.0	0.0021	11.2	0.282
4.0	0.0039	11.6	0.303
6.0	0.0058	11.6	0.318
7.0	0.0072	11.5	0.327
7.8	0.0084	11.2	0.351
8.3	0.0280	10.8	0.367
8.7	0.0530	10.1	0.394
9.2	0.0860	9.7	0.402
9.8	0.1450	9.6	0.420
10.4	0.2220		

5.35 A standard tensile specimen 12.7 mm in diameter was fitted with an extensometer with a 50 mm gage length and tested until failure. The force and extension measured during the test is presented in the table below. The diameter of the specimen at the fracture neck was 9.0 mm. Analyze the data using a spreadsheet and determine the following quantities.

    a.   Modulus of elasticity
    b.   Yield strength (0.2% offset)
    c.   Ultimate strength
    d.   Fracture stress
    e.   Strain at fracture
    f.   Percent elongation
    g.   Percent reduction in area.

Load (kN)	Extension (mm.)	Load (kN)	Extension (mm)
0	0	35.0	3.54
5.0	0.010	37.0	4.52
10.0	0.019	38.0	5.53
15.0	0.029	38.0	6.18
20.0	0.042	36.0	6.66
25.0	0.048	33.0	7.35
26.0	0.50	30.0	7.86
27.0	0.95	27.0	8.42
29.0	1.50	23.0	8.78
31.5	2.21	19.0	9.32

CHAPTER 6 PROBLEMS

6.1   Why is a truss used in constructing a bridge or a roof covering for a large structure?

6.2   What is the inherent geometric form that is used repeatedly in all truss designs? Why?

6.3   If a truss is fabricated from 37 members, determine the number of joints that will be required in constructing the truss.

6.4   Why is it necessary to employ an odd number of members in constructing a truss?

6.5   Prepare a sketch of a joint using a gusset plate for connecting four uniaxial bars in a truss structure. Identify all the components included in the joint that you have designed.

6.6   Why is it important to restrict the application of external loads on a truss to only the joints?

6.7   Consider the four-unit Howe truss shown in the figure to the right. If the external forces are as specified in this figure, use the method of joints to determine the forces in members AC, BC, CD and CE.

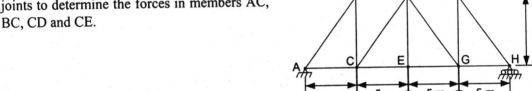

6.8   Consider the four-unit Howe truss shown in the figure to the left. If the external forces are as specified in this figure, use the method of joints to determine the forces in members AC, BC, CD and CE.

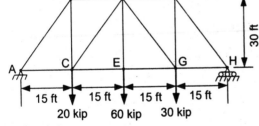

6.9   Consider the four-unit Howe truss shown in the figure to the right. If the external forces are as specified in this figure, use the method of joints to determine the forces in members GE, GD, GF and GH.

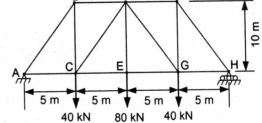

6.10 Consider the four-unit Howe truss shown in the figure to the right. If the external forces are as specified in this figure, use the method of joints to determine the forces in members AC, BC, CD and CE.

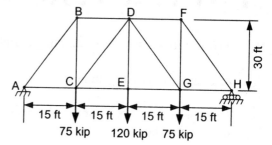

6.11 If the safety factor SF for each of the members in the truss of Problem 6.7 is specified as 3.0, determine the minimum required cross sectional area for members AC, BC, CD and CE. Note the yield strength S_y of the hot rolled structural steel used in fabricating the truss is 200 MPa.

6.12 If the safety factor SF for each of the members in the truss of Problem 6.8 is specified as 3.2, determine the minimum required cross sectional area for members AB, AC, BD and BC. Note the yield strength S_y of the hot rolled structural steel used in fabricating the truss is 30 ksi.

6.13 A scissors truss, illustrated in the figure to the right, is loaded with three forces at the joints B, C and D. If the truss is fabricated from 1018 A steel, determine the size of members AB, AF, BC and BF. The safety factor is specified as 3.0.

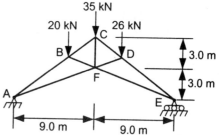

6.14 A scissors truss, illustrated in the figure to the left, is loaded with three forces at the joints B, C

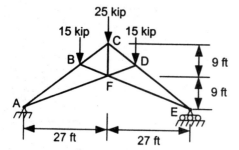

and D. If the truss is fabricated from 1020 HR steel, determine the size of members AB, AF, BC and BF. The safety factor is specified as 2.9.

6.15 For the Howe truss, shown in the figure to the right, use the method of sections to determine the forces in the members BD, CD and CE.

6.16 For the Howe truss, shown in the figure to the right, use the method of sections to determine the forces in the members DF, DG and EG.

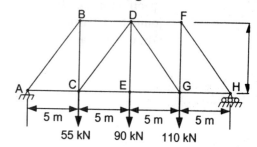

6.17 For the truss, defined in the figure to the right, use the method of sections to determine the forces in members DF, DB and DE. The external forces applied at joints C, E, G, I, and K are each equal to 15 kip.

6.18 If the cross sectional areas of all of the members in the truss defined in Problem 6.17 are equal to A = 2.3 in.², determine the safety factor for structural members DF, DB and DE. The yield strength of the steel used in the truss members is 35 ksi.

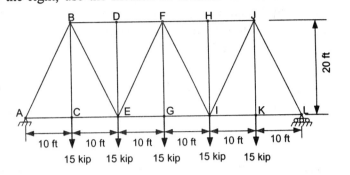

6.19 By inspection identify the zero-force members in the truss defined in Problem 6.17.

6.20 For the truss, defined in the figure to the right, use the method of joints to determine the forces in members DC, DE and DF.

6.21 If the cross sectional areas of all of the members in the truss defined in Problem 6.20 are equal to 3,000 mm², determine the safety factor for the members DC, DE and DF. Note the yield strength of the steel used in the truss is 310 MPa.

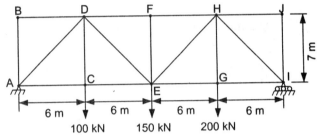

6.22 Use the method of sections with the truss, shown in the figure for Problem 6.20, to determine the forces in members FH, EH, and EG.

6.23 Let s/h be a variable in the truss structure defined in the figure to the left. Determine the forces in the members AC and AD as a function of the s/h ratio. We suggest you use a spreadsheet to perform the calculations. Consider the ratio s/h over the range from 0.5 to 2.0 varying in steps of 0.1. Prepare a graph of the results for the forces in members AC and AD as a function of s/h.

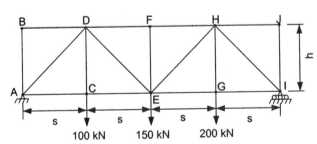

6.24 Determine the forces in members CE, DE and DF of the bowstring truss shown in the figure to the right. The span s = 12 ft.

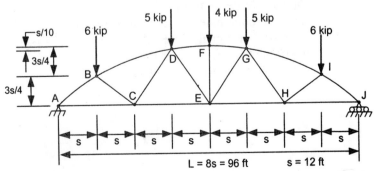

6.25 Determine the forces in members AC, CD, CE and DF of the inclined truss shown in the figure to the right.

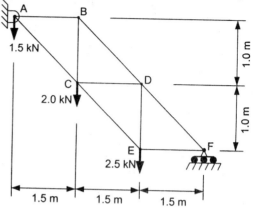

6.26 Determine the forces in members CD, CJ, and KJ of the truss shown in the figure below.

6.27 Determine the forces in members DE, EJ, and JI of the truss shown in the figure to the right.

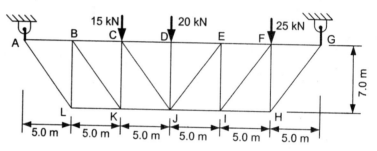

6.28 Determine the forces in members EF, IF, and IH, of the truss shown in the figure below.

6.29 Determine the forces in members BC, BK, and LK, of the truss shown in the figure to the right.

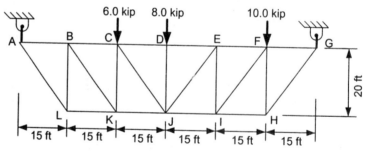

6.30 For the cantilever truss, shown in the figure to the right, determine the stress in member BC if its cross sectional area is 2.5 in². Also determine the safety factor for this truss member if it is fabricated from 1020 HR steel.

6.31 For the cantilever truss, shown in the figure to the right, determine the stress in member AB if its cross sectional area is 3.5 in². Also determine the safety factor for this truss member if it is fabricated from 1020 HR steel.

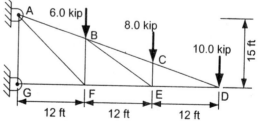

6.32 For the cantilever truss, shown in the figure to the right, determine the force in member GF. Specify the minimum cross sectional area if the member is to exhibit a safety factor of 3.0 relative to the yield strength of its material. This truss member is fabricated from 1020 HR steel.

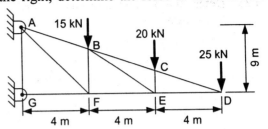

6.33 For the cantilever truss, shown in the figure to the right, determine the stress in member CD if its cross sectional area is 900 mm². Also determine the safety factor for this truss member if it is fabricated from 1020 HR steel.

6.34 For the cantilever truss shown in Problem 6.33, determine the force in member ED. Specify the minimum cross sectional area if this member is to exhibit a safety factor of 3.0 relative to the yield strength of its material. This truss member is fabricated from 1212 HR steel.

6.35 For the truss, presented in the figure below, determine the cross sectional area required for members AB, BD, DE, EF and CF if they are fabricated from 1020 HR steel and a safety factor of 3.2 is specified.

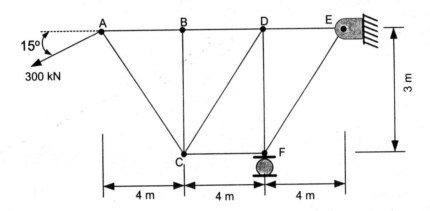

CHAPTER 7 PROBLEMS

7.1   The Patcenter in Princeton, NJ is a large open structure with a cable-stayed roof. In the figure presented below, one of the nine tubular steel masts that are uniformly spaced at 9 m intervals to support the roof structure is shown. In this design, the roof hangs from cables because the vertical side members are used only to prevent wind uplift. The tubular mast, 15 m high, is supported with a 9 m wide by 6 m high rectangular steel frame. Determine the force in the primary rod stay and the tubular steel masts if the uniformly distributed load on the roof is 70 kN/m. Assume the roof is pinned to the steel mast at point C. Hint: Refer to the Example 7.3 in the text for additional dimensions.

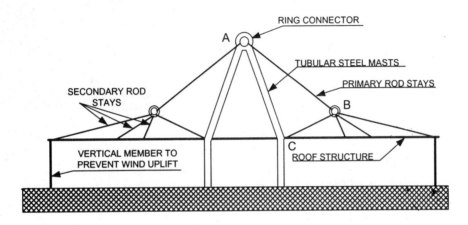

7.2   A new building similar to the Patcenter is under consideration by an architectural firm. They propose increasing the height of the mast from 15 m to 24 m to make the appearance of the structure more dramatic. If all of the other parameters are the same as given in Problem 7.1, prepare an analysis indicating the effect of this change.

7.3   A new building similar to the Patcenter is under consideration by an architectural firm. They propose a structure with the dimensions presented in the figure to the right. Determine the force in the primary rod stay AB and the tubular steel masts AC. Also find the reaction forces at point C.

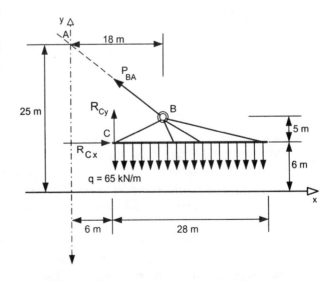

7.4 A space truss, shown in the figure to the right, is supported at point A with a ball and socket joint and with cables anchored at points B and D. A force F is applied to joint E. If F is represented in vector format as shown in the figure, determine the internal forces in the three members that intersect at joint E. Also specify the diameter of the rods, if the truss is fabricated from a 1018 A steel alloy. A safety factor of SF = 3.0 based on yield strength is specified by the building code. The truss dimensions are given in ft.

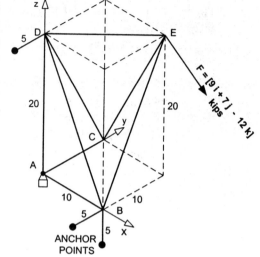

7.5 Determine the internal forces in members AB, DB and EB of the truss described in Problem 7.4.

7.6 Determine the internal forces in members AC, DC and EC of the truss described in Problem 7.4.

7.7 For the communications tower illustrated in the figure to the right, a wind force of 4,500 lb from the West is applied at a position $z = 300$ ft. The tower has a height H = 400 ft and a dead weight W = 50 kips. The anchor points on the ground plane are defined by: $x_1 = 100$ft, $x_2 = 115$ ft, $y_1 = 130$ ft and $y_2 = 145$ ft. The cables have a breaking strength of 26,600 lb and a ball and socket joint is used to support the tower at point O. Determine the loads in the four cables and the reaction forces at the base of the tower. Also determine the minimum safety factor.

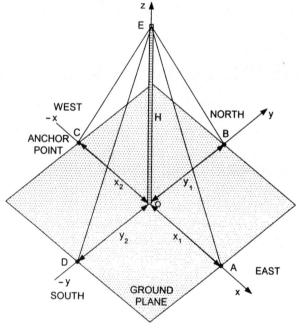

7.8 The anchor points on the ground plane for the communications tower shown in the figure to the right are changed to: $x_1 = 60$ m, $x_2 = 70$ m, $y_1 = 80$ m, $y_2 = 90$ m. The tower has a height of 120 m, a dead weight of 200kN and is subjected to a wind force of 12.6 kN from the East, that is applied at a position $z = 80$ m. The cables have a breaking strength of 130 kN, and the tower is supported by a ball and socket joint at point O. Determine the loads on the four cables and the reaction forces at the ball and socket joint. Also find the minimum safety factor.

7.9 If the steel used in fabricating the tower described in Problem 7.7 has a yield strength of 50 ksi, determine the cross sectional area required at the base of the tower. The safety factor for the tower is specified as 4.0. Does the wind loading condition affect the result?

7.10 If the steel used in fabricating the tower in Problem 7.8 has a yield strength of 360 MPa, determine the cross sectional area required at the base of the tower. The safety factor for the tower is specified as 5.0. Does the wind loading condition affect the result?

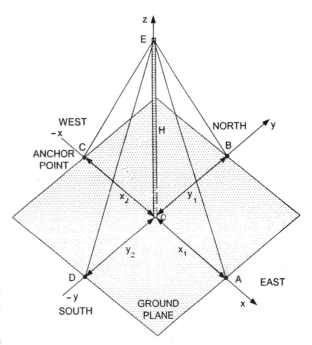

7.11 The safety factors for the communications tower described in Problems 7.7 and 7.8 are very large. Develop arguments supporting the use of relatively large safety factors in the design of communications towers. Also develop arguments for redesign with smaller diameter cable, smaller footprint and smaller cross sectional area at the base of the tower structure.

7.12 A severe ice storm strikes the communications tower coating all of the members with a thick layer of ice. The dead weight of the tower is increased from 200 kN to 320 kN. Determine the decrease in the safety factor for the conditions of Problem 7.8.

7.13 For the derrick shown in the figure below, determine the margin of safety for the three cables if the force F acting on the boom is 4.0 ton. The cable diameter is 0.5 in. for all three cables. The anchor points P and Q for two of the cables lie in the x-y plane with coordinates P= (12, −9, 0) ft and Q = (−12, − 9, 0) ft. The derrick pole is 15 ft high and the boom is 20 ft long. The derrick pole and the boom are separate components. The boom is attached to the pole at point R with a sleeve type bearing that enables the boom to rotate about the pole. However, the boom is constrained from sliding up or down the pole. The derrick pole is supported by a ball and socket joint at point O. Details of the bearing arrangement are also shown in the figure below. The cables have a specified breaking strength of 180 ksi.

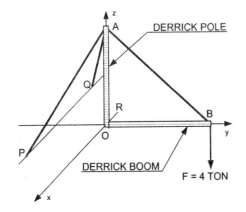

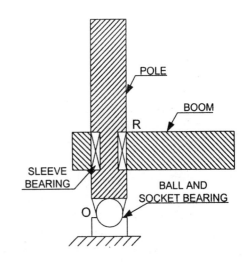

7.14 A tetrahedral space truss, shown in the figure to the right, supports a massive scoreboard and several sets of remotely controlled spotlights in an amphitheater. The base ABC of the space truss lies in the x-y plane (horizontal). The base triangle is connected to anchors in the roof beams by long cables. Determine the size of the solid round rods that are required to fabricate members AB, AC and AD of the space truss. The safety factor based on yield strength is specified as 6.0. 1020 HR steel is employed for all of the members and the weight of the scoreboard and spotlights is 20.0 kN.

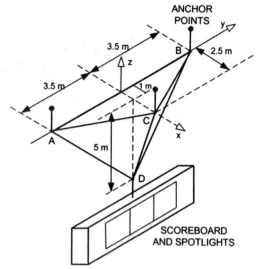

7.15 For the long horizontal boom of a construction crane, illustrated in the figure shown below, determine the internal forces in the members AB, BF, BK, FG, KL and FL. The load applied to the boom is F = 10 kip. Hint: See Example 7.8 in the text for additional details.

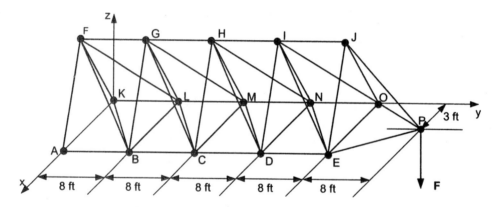

7.16 Repeat Problem 7.15 if the loading on the long horizontal boom is increased from 10 kip to 14 kip.

7.17 A hanging light assembly is positioned near the corner of a gymnasium as shown in the figure to the right. Determine the gage of the stainless steel wire required for the support of a light assembly weighing 1,100 N if a safety factor of 5.0 is specified. The wire is fabricated from 302A stainless steel. Points B and C are anchors that lie in the x-y plane and point D is an anchor that lies along the z-axis. Point A is not anchored; however it lies in the x-y plane.

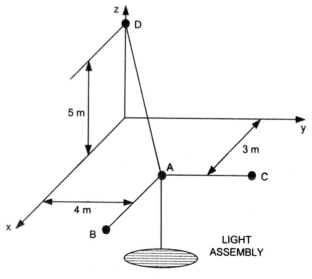

7.18 A hot air balloon, shown in the figure to the right, is moored to the ground with three cables that are anchored at points A, B and C. The coordinates (x, y, z) of the anchor points on the ground and on the basket are given in the figure. Determine the force exerted by each of the cables if the upward lift of the balloon is 950 lb. Assume that wind forces are negligible.

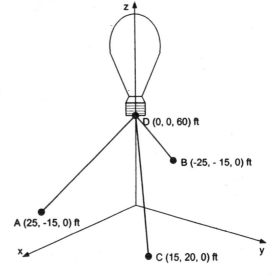

7.19 A circus cage, displayed in a large high ceiling auditorium, is supported above ground level by the three wires illustrated in the figure to the left. Determine the gage of the stainless steel wire required to support the cage weighing 2,500 lb if a safety factor of 3.5 is specified. The wire is fabricated from 4340 HR steel. The geometric parameters defining the assembly are listed in the table below. Points A, B and D locate the anchors for the cables.

7.20 A local firm has constructed a small crane consisting of a boom supported by two steel wires BC and DE as shown in the figure to the right. The boom is fixed to the supporting wall with a ball and socket joint at point A. The wires are anchored into the wall at points C and E. Each wire has a diameter of 0.125 in. and an ultimate tensile strength of 150 ksi. Determine the maximum weight that can be supported by the boom, the support reactions at point A and the forces in wires BC and DE.

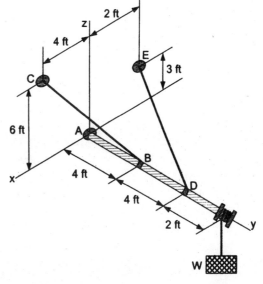

7.21 A hand operated lifting mechanism called a windless utilizes a crank to rotate a drum as shown
in the figure to the right. The shaft of the
mechanism is supported by a wall mounted ball
and socket joint at point A and a smooth journal
bearing at point B. The arm and handle of the
crank in the position shown is in the y-z plane.
For this position of the crank handle, determine
the force F required to hold a weight of 500 kN
in equilibrium. Also determine the reactions at
the ball and socket joint and the journal bearing.

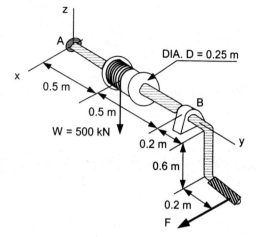

7.22 Repeat Problem 7.21 if the crank is rotated
clockwise 90 degrees so that the arm and handle
of the crank is in the x-y plane and the force F is
acting in the positive z direction. For this
position of the crank handle, determine the force
F required to hold the weight of 500 kN in equilibrium. Also determine the reactions at the ball
and socket joint and the journal bearing.

7.23 Repeat Problem 7.21 if the crank is rotated clockwise 180 degrees so that the handle of the
crank is in the y-z plane and the force F is acting in the negative x direction. For this position of
the crank handle determine, the force F required to hold the weight of 500 kN in equilibrium.
Also determine the reactions at the ball and socket joint and the journal bearing.

7.24 A cross-like base and a short pole, as shown in the figure to the right, support a tabletop with a
uniform thickness. Determine the largest weight
W that can be applied to the table if it is placed at
points A, B or C. The tabletop weighs 120 N/m².
The dimensions of the tabletop and the base are
given in the figure.

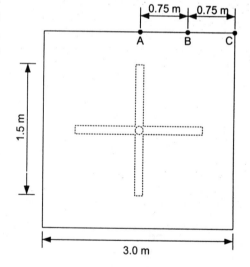

CHAPTER 8 PROBLEMS

8.1   For the hoist shown in the figure to the right, determine all of the external forces on member CFG as point D is moved along member AE so that the angle θ varies from 0 to 45°. We suggest that you use a spreadsheet in preparing this solution. Note the angle β = 60° and F = 580 N. The dimensions are: CF = 200 mm, FG = 320 mm and the pulley radius r = 10 mm.

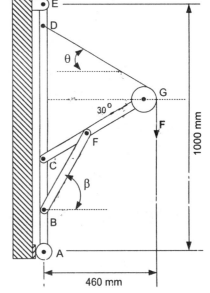

8.2   For the hoist shown in the figure to the right, determine all of the external forces on member CFG as point B is moved along member AE so that the angle β varies from 35 to 75°. We suggest that you use a spreadsheet in preparing this solution. Note the angle θ = 30° and F = 240 N. The dimensions CF = 200 mm, FG = 320 mm and the pulley radius r = 10 mm.

8.3   Prepare a FBD for the entire rectangular frame, shown in the figure to the left, if it is subjected to a concentrated force of 12.4 kN located a distance of 3.0 m from the left end of the horizontal member.

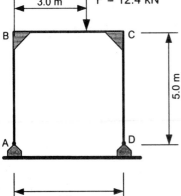

8.4   Determine the force F required to develop a pressure p = 2.1 MPa in the cylinder of the pump shown in the figure to the right. The piston area for the pump is 300 mm².

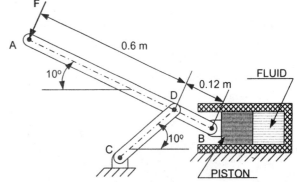

8.5   Determine the reaction force at point G and the mechanical advantage of the toggle mechanism illustrated in the figure to the right. A force F of 300 N is applied to the lever at point A. The contact at point G is made with a roller, and the links are connected with pins inserted at points B, C, D and E.

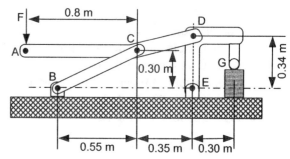

8.6   Determine the reaction force at point G and the mechanical advantage of the toggle mechanism illustrated in the figure to the right. The input force F is now 500 N. The contact at point G is made with a roller and the links are connected with pins inserted at points B, C, D and E.

8.7   Your manager believes that the dimension (e − d) for the toggle mechanism shown in the figure to the left is a critical design parameter. She asks you to determine the mechanical advantage of this mechanism if the dimension d is fixed and the dimension e is modified so that (e − d) varies from 5 mm to 50 mm. You may consider using a spreadsheet for this analysis.

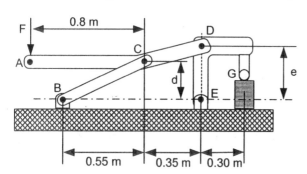

8.8   If the spring constant of the block under point G is 2.0 kN/mm, determine the vertical displacement of point G in Problem 8.5.

8.9   If the spring constant of the block under point G is 2.0 kN/mm, determine the work output of the toggle mechanism for the conditions described in Problem 8.5.

8.10   If the spring constant of the block under point G is 3.8 kN/mm, determine the vertical displacement of point G in Problem 8.6.

8.11   If the spring constant of the block under point G is 3.8 kN/mm, determine the work output of the toggle mechanism for the conditions described in Problem 8.6.

8.12   For the compaction press, shown in the figure to the right, determine the compressive force developed by the sliding platen if a force of 5 ton is applied to the toggle mechanism at point B. Also determine the mechanical advantage.

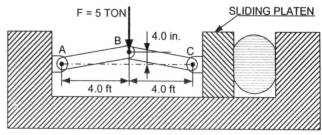

COMPACTION PRESS

8.13 A pair of pliers clamps a small diameter rubber cylinder as shown in the figure to the right. If opposing forces of 30 lb are applied to the handles of the pliers, determine the reaction forces acting on the cylinder. Also determine the work performed if each handle moves though a distance d = 0.14 in. Finally, determine the amount that the cylinder is squeezed by the application of the forces. The dimensions of the pliers are a = 1.0 in. and b = 5.5 in.

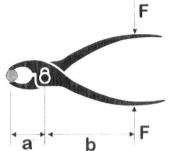

8.14 For the frame, shown in the figure to the left, determine the forces and moment at the fixed support at A, the forces at pin C and the force in link BD.

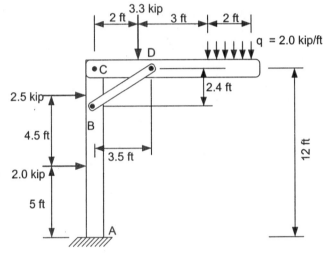

8.15 For the frame, illustrated in the figure to the right, determine all the forces acting on member ABCDE. Also determine the internal force in member AF. The attached weight W is 4.8 kN and the pulley radius is 150 mm. The dimensions in the figure are given in mm.

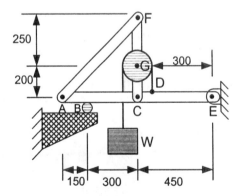

8.16 For the frame illustrated in the figure to the right, determine all the forces acting on member ABCDE. Also determine the internal force in member AF. The attached weight W is 4.8 kN and the pulley radius is 100 mm. The dimensions in the figure are given in mm.

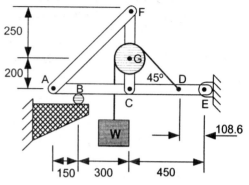

8.17 For the arm of the backhoe, shown in the figure below, determine the force that must be exerted by the hydraulic actuator AC to maintain the bucket loaded with a weight of 1.0 kN in equilibrium. Also find the forces in links BC and CE and the force acting on pin D. The bucket and the crosshatched appendage are welded together.

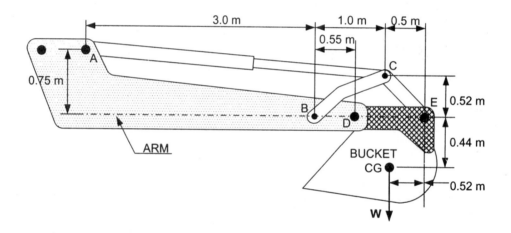

8.18 For the "Loadall" arm, presented in the figure below, determine the forces exerted by the hidden actuator and the visible actuator BC to maintain equilibrium. Also determine the forces acting at pins A and D. The arm weighs 1.5 kN and its center of gravity is located 1.8 m to the right of point A. All of the dimensions are given in meters.

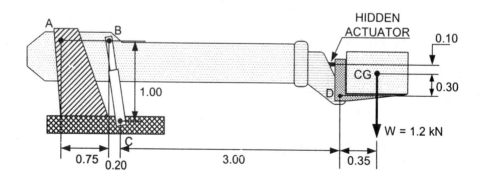

8.19 Levers are well known machines used for either amplifying or attenuating forces. When used in scales to weigh heavy objects, the levers are often arranged to compound the attenuation. Such an arrangement is illustrated in the figure below. If the pins are frictionless, show that the relation between the known scale weight w and the unknown scale weight W is:

$$W = w\left[\frac{(a+b+c+d-x)(a+b+c)b}{aec}\right]$$

8.20 Select dimensions a through e if the scale, illustrated in the figure above, is to measure a weight W = 5.0 kN with a small sliding weight w = 10 N. The value of x = 200 mm is fixed in this design analysis.

8.21 The horizontal boom of a construction crane is counter balanced with a 15 kip weight that is centered at point B. A cable BFE anchored at points B and E supports the horizontal boom. The cable is maintained at a constant tension over its length with a small pulley located at point F. The horizontal boom is attached to the tower with a pin located at point C. The crane is lifting a load W = 10 kip from a hoist located at point D. Determine all of the forces acting on the horizontal boom, the cable tension, all of the reactions at point A and the force acting on the pin at point F. The weight of the boom is 5.0 kip and the weight of the tower member is 4.0 kip.

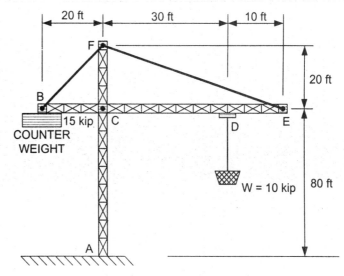

CHAPTER 9 PROBLEMS

9.1 Write an engineering brief describing the factors affecting friction forces acting between two bodies.

9.2 Write an engineering brief explaining the reasons for the difference between static and dynamic friction coefficients.

9.3 Derive Eq. (9.2).

9.4 Explain the conditions that must exist before you apply the equation $(F_f)_{max} = \mu N$.

9.5 Suppose you develop a fixture consisting of an inclined plane and a block of some solid material to measure the material's coefficient of friction. You measure and record the angle of the inclined plane at the instant motion is initiated. Determine the average value of the friction coefficient from the five separate tests and its range using the data given in the table below.

Problem No.	Angle	Angle	Angle	Angle	Angle
9.5a	22.0°	22.2°	24.3°	21.8°	22.5°
9.5b	24.0°	23.2°	26.3°	24.8°	23.5°
9.5c	32.0°	31.2°	33.3°	32.8°	32.7°
9.5d	37.0°	38.2°	37.9°	39.0°	39.2°

9.6 Derive Eq. (9.7).

9.7 Determine the force F required to move the sled, illustrated in the figure to the right, if the coefficient of friction between the sled and the surface is $\mu = 0.28$.

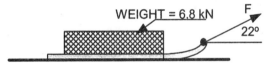

WEIGHT = 6.8 kN
F
22°

9.8 Determine the force required to slide a box, illustrated in the figure to the left, across a level floor if the force is applied at an angle $\alpha = 10°$. The coefficient of friction between the box and the floor is $\mu = 0.25$.

F
α
450 lb

9.9 A worker attempts to slide a wooden crate along a level concrete floor in a warehouse as shown in the figure to the right. The weight of the crate is 80 lb and the worker weighs 120 lb. The coefficient of friction between the crate and the floor $\mu_c = 0.55$ and between the worker and the floor $\mu_w = 0.70$. Note that the crate is short and the worker must lean over to apply a force at an angle of 30° relative to the floor. Is it possible for the worker to move the crate? Justify your answer by showing all the relevant calculations.

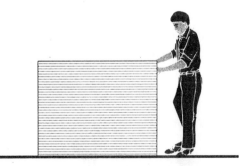

9.10 It has snowed on the day of your final examination for this course, and you find the roads covered with a coating of ice and snow. In your attempt to drive to the University, you cannot negotiate a hill with a grade of 4°. Estimate the coefficient of friction between the road and the tire for your instructor, as you explain the reason for missing the final examination.

9.11 Two blocks rest on a surface as illustrated in the figure to the right. If F_2 = 400 lb, determine the maximum force F_1 that can be applied before one of the blocks moves. The coefficient of friction between the two blocks is 0.28 and between block B and the surface it is 0.22. Block A weighs 1,200 lb and Block B weighs 2,000 lb.

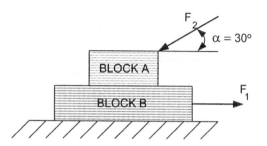

9.12 Determine if the ladder, shown in the figure to the left, is stable or not. The man on the ladder weighs 145 lb and the ladder weighs 52 lb. The coefficient of friction at the base of the ladder is 0.30 and the coefficient of friction between the ladder and the wall is 0.15.

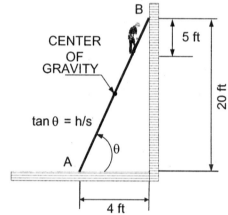

9.13 Determine the force required to move the crate, shown in the figure to the right, and indicate whether it will slide or tip. The crate weighs 720N and the coefficient of friction between it and the floor is μ = 0.42.

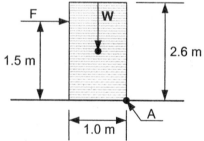

9.14 A cylindrical assembly is pulled by a force F to roll up and over a step as shown in the figure to the left. The cylinder weighs 420 N, and the coefficient of friction between the cylinder and the surfaces at points A and B is 0.43. The radii r_1 = 0.5 m and r_2 = 1.0 m. Determine if the cylinder will roll up and over the step or slip and remain in the corner. Show all of your work and justify your answer.

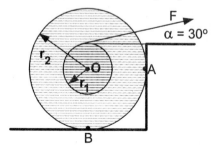

9.15 A wedge inserted at point A is used to lift one end of the machine bed shown in the figure to the right. If the wedge has an included angle $\alpha = 12°$, determine the force necessary to drive the wedge if the coefficient of friction with both surfaces $\mu = 0.35$, and the weight per unit area of the machine bed is 4 kN/m².

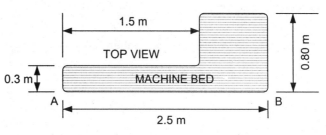

9.16 Reconsider Problem 9.15; however, the wedge is to lift the machine bed from point B and the angle of the wedge is reduced from 12° to 6°.

9.17 Determine the machine coefficient for the wedge, which measures its effectiveness as a simple lifting machine for the results obtained in Problem 9.15. Discuss the parameters leading to the best machine coefficient. Recall, work is defined as **F • d**.

9.18 If the machine bed in Problem 9.15 is lifted through a distance of 60 mm, determine the work expended in driving the wedge. Also determine the work accomplished in lifting the weight of the machine bed. Recall that work is defined as **F • d**.

9.19 If the machine bed in Problem 9.16 is lifted through a distance of 30 mm, determine the work expended in driving the wedge. Also determine the work accomplished in lifting the weight of the machine bed. Recall that work is defined as **F • d**.

9.20 For the conditions described in Problem 9.15, determine if the wedge will hold when the driving force is removed. Also determine the force F_R required to remove the wedge.

9.21 For the conditions described in Problem 9.16, determine if the wedge will hold when the driving force is removed. Also determine the force F_R required to remove the wedge.

9.22 You are lifting one corner of a car with a screw jack. The car weighs 3800lb, and the screw has a helix angle of 4°. The screw is lubricated with a thick, heavy grease and the coefficient of friction between the screw and the nut is $\mu = 0.11$. Determine the torque necessary to turn the screw.

9.23 You are lifting one corner of a heavy plate of steel with a screw jack. The plate weighs 10,000 N, and the screw has a helix angle of 3°. The screw is lubricated with a thick, heavy grease and the coefficient of friction between the screw and the nut is $\mu = 0.11$. Determine the torque necessary to turn the screw.

9.24 You are lifting one corner of a truck with a screw jack. The truck weighs 25 kN, and the screw has a helix angle of 4.5°. The screw is lubricated with a heavy grease and the coefficient of friction between the screw and the nut is $\mu = 0.11$. Determine the torque necessary to turn the screw.

9.25 Determine the locking angle α for helical threads if the coefficient of friction is 0.11.

9.26 Write an engineering brief describing the dangers associated with using a screw with a helical angle of more than 6° in the design of a jack to be used in lifting heavy weights. The screw is made from steel and the nut is made from brass.

9.27 A barrel with a diameter of 0.7 m and weighing 500 N is being rolled up an inclined ramp as shown in the figure to the right. The ramp weighs 250 N. If the coefficient of friction is 0.24 between all contacting surfaces, determine the distance x through which the barrel can be rolled before slippage occurs.

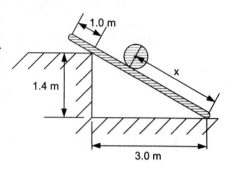

9.28 Block B with a mass m_B = 19 kg sits on top of a larger block A with a mass m_A = 32 kg. A horizontal force F is applied to block B as shown in the figure to the left. If the coefficient of friction between all of the surfaces is 0.28, determine the force F required to initiate motion. Describe the motion that occurs.

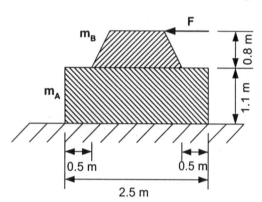

9.29 The links in the toggle mechanism, depicted in the figure to the right, are sufficiently light in weight to be neglected. The pins are frictionless. The coefficient of friction between the block and the surface is 0.34, and the block weighs 3.8 kN. Determine the angle θ that permits the application of the largest force F for which motion of the block does not occur.

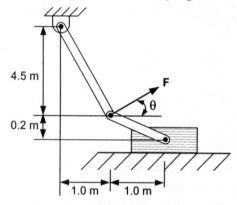

9.30 A carton weighing W_A = 85 lb has tipped and is resting against another carton weighing W_B = 125 as shown in the figure to the left. The coefficient of friction between the floor and the first (tipped) carton is μ_A = 0.45 and the coefficient of friction between the floor and the second (upright) carton is μ_B = 0.40. Determine if the boxes are in equilibrium in the position shown. Neglect the effect of friction between the cartons at the contact point.

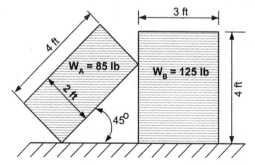

9.31 A crate with a weight W = 2.5 kN is connected to two opposing weights W_1 and W_2 by thin cables that pass over frictionless pulleys as shown in the figure to the right. Determine the minimum and maximum values for the weight W_1 so that motion does not occur. The coefficient of friction between the crate and the floor is 0.28.

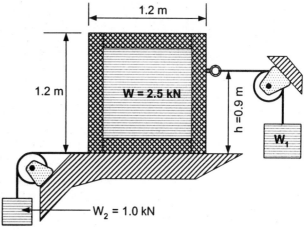

9.32 Consider the crate, shown in the figure to the right, that is subjected to the conditions given in Problem 9.31 except for the height h. Determine the product of h times the weight W_1 that produces motion by slipping and tipping simultaneously.

CHAPTER 10 PROBLEMS

10.1 Draw three different shapes of cross sectional areas that exhibit symmetry about the vertical or y axis.

10.2 For the beam shown in the figure to the right, construct the shear and bending moment diagrams and write the equations for the bending moment as a function of x as it varies from zero to L.

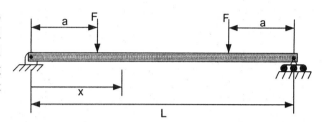

10.3 In a manufacturing process, an aluminum sheet with a thickness of 1.2 mm is passed around a roller with a 450 mm diameter. Determine the maximum strain in the sheet. Also determine the maximum stress.

10.4 A beam with a rectangular cross section has a depth b = 80 mm and a height h = 600 mm. If the yield strength of the beam's material is 200MPa, determine the maximum moment that the beam can support before it yields if the margin of safety is 100%.

10.5 Three planks with dimensions of 2.0 by 8.0 inches are adhesively bonded together to form a timber I beam as shown in the figure to the right. Determine the stresses due to bending at points A, B, C and D. The bending moment applied to the beam is 27,000 ft-lb.

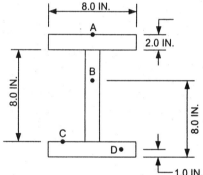

10.6 A WT6 × 60 structural T section is employed as a cantilever beam. The beam is fabricated from steel that has a yield strength of 40 ksi in tension and 85 ksi in compression. Bending occurs about the z-axis and a safety factor of at least 2.8 is required. Determine the largest positive bending moment that the beam can withstand if all of the requirements are satisfied. Also determine the maximum negative bending moment that the beam can withstand.

10.7 For the beam section—W406 × 100, calculate values of I_z and Z relative to the Z-Z axis from data given in Appendix D. Compare your values with those listed in the table.

10.8 Consider the flange and web section, defined in the figure to the left. If $b_w = xb$ and $h_w = yh$, determine the effectiveness **SE** as x varies from 0.05 to 0.3 and y varies from 0.6 to 0.95. It is suggested that you perform these calculations on a spreadsheet and prepare a graph displaying your results.

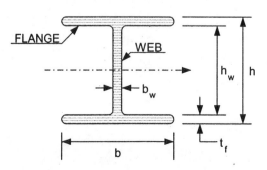

10.9 Two steel plates 9.0 in. by 0.75 in. are welded to the flanges of an S18 × 70 American Standard I-beam as shown in the figure to the right. Calculate the moment of inertia about each of the axes of symmetry.

10.10 If the thickness of the two plates were increased from 0.75 in. to 1.25 in. for the modified I-beam shown in the figure to the right, determine the moment of inertia about each of the axes of symmetry.

10.11 For the modified I-beam in the figure to the right with the 0.75 in. thick plates, determine the added capacity of the beam in supporting bending moments about the horizontal axis of symmetry.

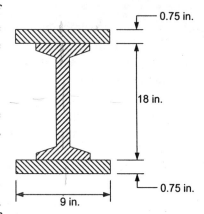

10.12 Four different beams with various end conditions are shown in the figure below. Construct FBDs of each beam showing the forces and/or moments at each support.

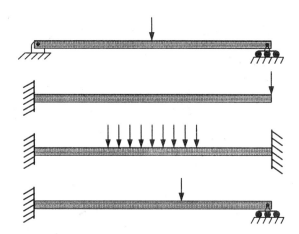

10.13 Write equations for the shear and bending moments as a function of position x across the length of the beam as shown in the figure to the right. Also construct the shear and bending moment diagrams.

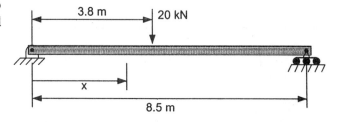

10.14 Write equations for the shear and bending moments as a function of position across the beam described in the figure to the left. Also construct the shear and bending moment diagrams.

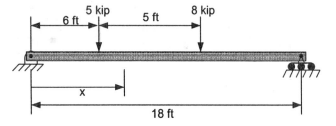

10.15 Write equations for the shear and bending moments as a function of position x across the beam described in the figure to the right.
Also construct the shear and bending moment diagrams.

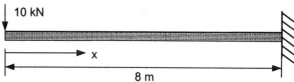

10.16 Write equations for the shear and bending moments as a function of position x across the beam described in the figure to the left.
Also construct the shear and bending moment diagrams.

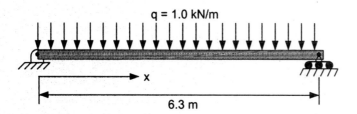

10.17 Write equations for the shear and bending moments as a function of position x across the beam described in the figure to the right.
Also construct the shear and bending moment diagrams.

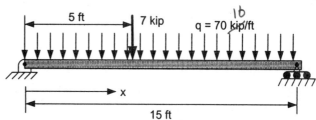

10.18 Write equations for the shear and bending moments as a function of position x across the beam described in the figure to the left. Also construct the shear and bending moment diagrams.

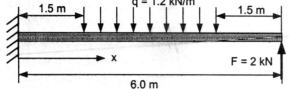

10.19 Select a WF section for the steel beam shown in the figure to the right. The margin of safety (MOS) is 200% and the yield strength of the material is 320 MPa.

10.20 Select an American Standard section for the steel beam shown in the figure to the right. The margin of safety (MOS) is 150% is and the yield strength of the material is 350 MPa.

10.21 Select a structural tee section for the steel beam shown in the figure to the right. The margin of safety (MOS) is 200% and the yield strength of the material is 300MPa.

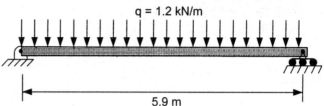

10.22 For the beam, described in the figure to the left, determine the maximum shear stress τ if the cross section of the beam is rectangular with a height h = 4.0 in. and a depth b = 1.0 in.

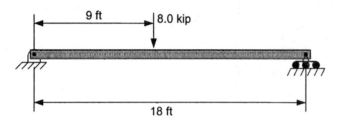

10.23 A circular shaft with a diameter of 50 mm is subjected to transverse loading from a gear set that produces a maximum shear force of 25 kN on the shaft. Determine the maximum shear stress τ_{Max} in the shaft due to this transverse load.

10.24 A timber company has designed a series of box beams with the geometry presented in the figure to the right. The beams are constructed by bonding the top and bottom boards to the edges of the vertical members at the four interfaces. Determine the maximum shear stress τ_{Max} and the shear stress at the adhesive joints. The vertical shear force V is 5.0 kN.

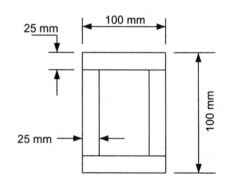

10.25 A timber company has designed a series of unsymmetrical T-sections to serve as beams as indicated in the figure to the left. The beams are constructed by bonding the top and bottom flanges to the edges of the vertical web at the two interfaces. Determine the maximum shear stress τ_{Max} and the shear stress at the adhesive joints. The vertical shear force V is 1,200 lb.

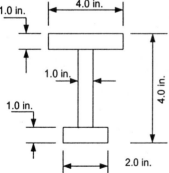

10.26 A timber company produces a series of I-beams fabricated from three pieces of timber as shown in the figure to the right. The flange of the I-beam is fastened to the web with spikes that are each capable of resisting a shear force of 960 lb. The beams are 12 ft long and are simply supported. They carry a concentrated force of 3,500 lb at midspan. Determine the maximum spacing of the spikes needed to transmit the shear from the web to the flange of the beam.

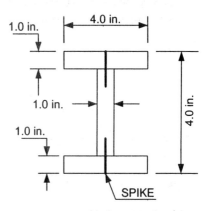

10.27 Derive Eq. (10.29).

10.28 Consider a simply-supported, foam core, metal cover plate composite beam as illustrated in the

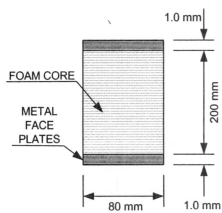

figure to the left. If the beam is loaded with a uniformly distributed load, determine the magnitude of q_Y that produces yielding. Aluminum cover plates are adhesively bonded to a polystyrene foam core. The beam is 2.5 m long and the yield strength of the aluminum is 280 MPa.

10.29 Design a portable, one-way, bridge beam with a single structural member consisting of metal cover plates and foam core. The bridge is to support the weight of four people uniformly distributed over the span of the bridge beam. Assume that the average weight of a typical person is 700 N. The bridge beam spans a ravine 5 m wide.

10.30 A cantilever beam 1.5 long is subjected to a concentrated force of 1,200 N applied at its free end. The beam has a rectangular cross section with a width of 90 mm. Determine the height of the foam h_f and the thickness of the cover plates t_m if the design stress in these plates is limited to 160 MPa. In this determination, recall that the standard thickness of aluminum sheet stock from which the cover plates are cut is 0.016, 0.020, 0.025, 0.032, 0.040, 0.050, 0.063, and 0.090 in.

10.31 A beam fabricated from western white pine, with dimensions shown in the figure to the right, has been reinforced by bonding steel strips on its top and bottom surfaces to form a composite beam. If the design stress is limited to 1.5 ksi in the wood and 30 ksi in the steel, determine the maximum bending moment if the beam bends about its horizontal axis of symmetry.

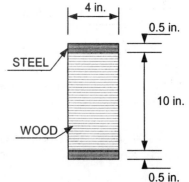

10.32 For the wood and steel composite beam described in Problem 10.31, determine the maximum bending moment if the beam is bent about its vertical axis of symmetry.

10.33 Derive Eqs. (10.32) and (10.35).

10.34 A concrete floor slab is reinforced with 12 steel rebars each with a diameter of ½ in. as illustrated in the figure below. The slab is 24 ft long and is simply supported when installed. If a uniformly distributed load q is applied to the floor, determine its maximum magnitude if the stress in the steel is limited to 20 ksi and in the concrete to 1,500 psi. Note the modulus of elasticity of the steel and the concrete is 30×10^6 and 2.8×10^6 psi, respectively. Comment on the balance of the design of this reinforced concrete slab.

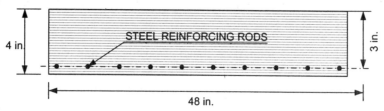

10.35 A concrete beam is reinforced with 8 steel rebars each with a ½ in. diameter as shown in the figure to the right. The beam is 20 ft long and simply supported. If it supports three concentrated forces each 6.0 kip positioned at L/4, L/2 and 3L/4, determine the margin of safety for the beam. The yield strength of the rebar is 36 ksi and the compressive strength of concrete is 4 ksi. The modulus of elasticity of the steel and the concrete is 30×10^6 and 2.8×10^6 psi, respectively.

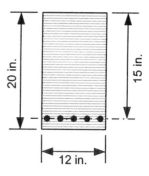

10.36 A high strength concrete beam is reinforced with 3 steel rods each with an area of 115 mm² located at positions shown in the figure to the left. For an applied bending moment of M = 5 kN-m, determine the maximum stress in the steel rebars and the concrete.

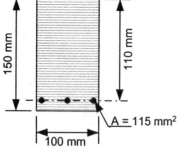

10.37 Discuss the advantage of using steel reinforced concrete beams as structural components in a building in the event of a severe fire. In your discussion, discuss the effect of prolonged exposure to flames on the strength of steel.

10.38 Redesign the steel reinforced concrete floor slab in Example 10.16 to improve the balance between the stresses in the steel and concrete.

10.39 Redesign the steel reinforced concrete beam in Example 10.17 to improve the balance in the margin of safety between the steel and concrete.

10.40 A beam with a rectangular cross section, defined in the figure to the right, is fabricated from steel with $S_y = 320$ MPa. Determine the yield moment M_Y.

10.41 A beam with a rectangular cross section, defined in the Problem 10.40, is fabricated from steel with $S_y = 320$ MPa. Determine the moment M_p required to drive the elastic-plastic interface to the position $y_p = h/4$.

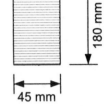

180 mm

45 mm

10.42 For a beam with a rectangular cross section, show that $M_P = M_Y$ when $y_p = h/2$. You may wish to examine Eq. (10.40) in developing this solution.

10.43 Determine the ratio of M_L/M_Y for the web and flange section, defined in the figure to the left, if $b_w = xb$ and $h_w = yh$. Let x vary from 0.05 to 0.3 and y vary from 0.6 to 0.95. It is suggested that you perform these calculations on a spreadsheet and prepare a graph displaying your results.

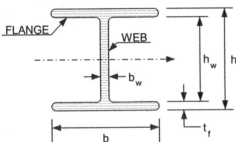

10.44 A notched cantilever beam, illustrated in the figures below, is notched at position x = 3.8 ft. Determine the maximum tensile stress on the beam.

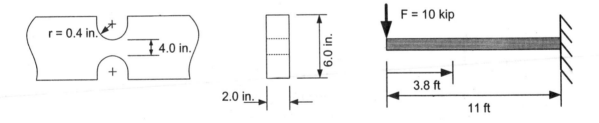

10.45 An architect has proposed a series of stepped roof beams to provide a visual impact for a public building with high pedestrian traffic. A representative beam with its loading is presented in the figures below. The beam's depth is 60 mm, and its yield strength is 240 MPa. As the city's structural engineer, you have the responsibility to approve or disapprove the design. Justify your decision.

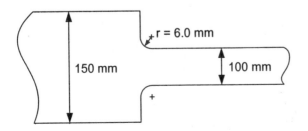

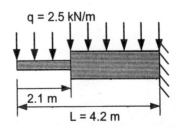

10.46 A shaft in a drive train, illustrated in the figures below, is subjected to bending moments due to a transverse force of 3.5 kN imposed by a gear set. Determine the maximum stresses at the fillet due to the gear load.

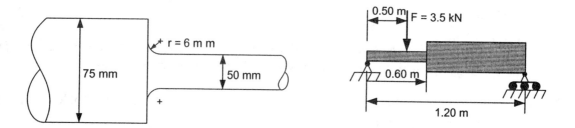

10.47 A shaft in a turbine engine is used as a conduit to transport coolant to the root of the turbine blades. The coolant enters and exits the shaft through small holes as shown in the figure below. Determine the maximum stress adjacent to the hole if the shaft supports a bending moment M = 400 N-m at the location of the hole.

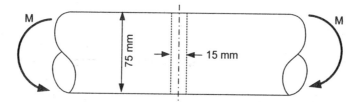

CHAPTER 11 PROBLEMS

11.1 A shaft is employed to transmit 700 W of power from an electric motor to an appliance. Determine the torque imposed on the shaft if the motor is operated at the angular velocity of 3600 RPM.

11.2 A shaft with a circular cross section is twisted through an angle $\phi = 3.1°$. If the diameter of the shaft is 15 mm and its length is 1.3 m, determine the maximum shearing strain γ.

11.3 Determine the maximum shearing stress τ_{max} for the conditions described in Problem 11.2 if the shaft is fabricated from steel.

11.4 A shaft with a circular cross section is subjected to a torque of 120 ft-lb. If the shaft's diameter is 0.750 in and its length is 15 in., determine the maximum shearing stress τ_{MAX}.

11.5 A shaft fabricated from 302 A stainless steel transmits 1.5 kW of power while rotating at an angular velocity of 1180 RPM. Specify a standard size round for the diameter of the shaft if a safety factor of 2.9 with respect to yielding is specified.

11.6 A hollow tube is employed in the design of a shaft to transmit 1.5 kW of power at 1180 RPM with a safety factor of 2.7 with respect to yielding. Determine the outside diameter d_o of the tube if its wall thickness is 0.05 d_o. The tube is fabricated from 1020 HR steel.

11.7 Clearly a hollow tube is structurally more efficient than a solid rod as a shaft for transmitting power. Cite several reasons for the limited usage of tubes to transmit power in actual practice.

11.8 Determine the section modulus of a circular tube with an outside diameter of 120 mm and an inside diameter of 90 mm.

11.9 Determine the structural efficiency *SE* of the shaft described in Problem 11.8.

11.10 For the inclined surface, defined in the figure to the right, determine the stresses τ_{nt} and σ_n on the inclined plane. The shear stress $\tau_{xy} = 6.5$ ksi and the angle $\theta = 22°$.

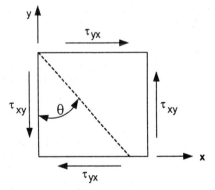

11.11 For the stress $\tau_{xy} = 75$ MPa specified in the figure to the right, prepare a graph of τ_{nt} and σ_n as θ varies from zero to 360°. Define the principal planes and the planes on which the shear stress is a maximum.

11.12 A solid circular shaft with a diameter of 1.0 in. is transmitting 7.0 HP from an electric motor to an air compressor. If the motor is operating at 1800 RPM, determine the maximum normal stress acting on the shaft. Also determine the plane upon which this maximum normal stress acts.

11.13 A steel shaft with a circular cross section is subjected to a torque of 120 ft-lb. If the shaft's diameter is 0.75 in. and its length is 15.0 in., determine its angle of twist ϕ.

11.14 A line shaft, shown in the figure below, is fabricated from high-strength steel and incorporates five spur gears. The gear A provides the input torque T_A and gears B, C, D and E each remove torque from the shaft. Determine the angle of twist for the entire shaft AE and between gears at A and C. The shaft diameters and lengths are given in the table below. The torque at each gear is also listed in this table. Neglect the effect of gear deformation on the angle of twist. Also determine the maximum shear stress in the shaft. Where does this shear stress occur?

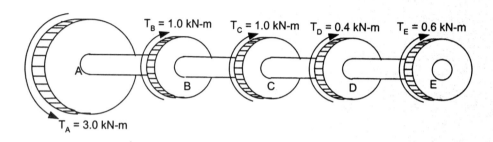

Prob. No.	d_{A-B}	d_{B-C}	d_{C-D}	d_{D-E}	L_{A-B}	L_{B-C}	L_{C-D}	L_{D-E}
11.14	40 mm	35 mm	30 mm	25 mm	0.4 m	0.4 m	0.5 m	0.7 m

11.15 A line shaft, shown in the figure below, is fabricated from high-strength steel and incorporates five spur gears. The gears A and E provide input torques T_A and T_E. Gears B, C, and D each remove torque from the shaft. Determine the angle of twist for the entire shaft AE and between gears at A and C. The shaft diameters and lengths are given in the table below. The torque at each gear is also listed in this table. Neglect the effect of gear deformation on the angle of twist. Also determine the maximum shear stress in the shaft. Where does this shear stress occur?

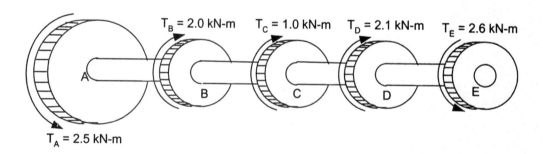

Prob. No.	d_{A-B}	d_{B-C}	d_{C-D}	d_{D-E}	L_{A-B}	L_{B-C}	L_{C-D}	L_{D-E}
11.15	40 mm	35 mm	30 mm	25 mm	0.4 m	0.4 m	0.5 m	0.7 m

11.16 A steel torsion bar is incorporated into the chassis of an automobile to serve as a suspension spring. If the bar is fabricated from steel rod with a diameter of 15 mm and a length of 1.1 m, determine its spring rate. Also determine its angle of twist if it stores a strain energy of $E = 6.0$ N-m as the automobile traverses a bump.

11.17 Determine the nominal and maximum shear stresses for a shouldered shaft shown below with a diameter D = 1.5 in. and a diameter d = 1.25 in. subjected to a torque of 500 ft-lb. The fillet radius at the shoulder of the shaft is 0.25 in.

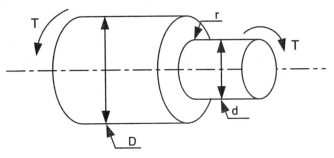

11.18 A shouldered shaft is designed such that the maximum shear stress τ_{Max} cannot exceed a critical value of 50 MPa. The shouldered shaft shown above and to the right has a diameter D = 40 mm and a diameter d = 30 mm. The fillet radius at the shoulder of the shaft is 6 mm. Determine the largest torque that the shaft can support.

11.19 The electric motor shown in the figure to the right has a power rating of 5.0 kW and operates at 1180 RPM. It powers a gear drive with a variable load, which sometimes requires the motor to operate at its maximum rating. The shaft is supported by two bearings— one on each side of the gear drive. The shaft from the motor has a diameter d and the shaft at the gear has a diameter D. The radius at the transition is specified as r/d = 0.15 and the diameter mismatch ratio is specified as D/d = 1.33. Select a material for the shaft and determine the diameters d and D if the maximum shear stress in the material is not to exceed 1/4 of the yield strength of the material in shear. Another constraint on the design is the angle of twist ϕ, which is not to exceed 2.0° measured from the motor to the gear drive.

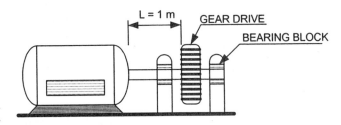

11.20 The electric motor, shown in the figure below, has a power rating of 20.0 HP and operates at 1780 RPM. It powers a drive shaft with three gears. When the motor is operating at rated capacity, gear A removes 1/2 of the power developed by the motor, while gear B and gear C remove 1/3 and 1/6 of the power, respectively. The shaft is supported by four bearings as shown in the figure. The shaft from the motor has a diameter d but it is shouldered with a diameter D at each of the three gear locations. The radius at the transition is specified as r/d = 0.20 and the diameter mismatch ratio is specified as D/d = 2.00. If an alloy steel with a yield strength in shear of 150 ksi is used to fabricate the shaft, determine the diameters d and D if the maximum shear stress in the material is not to exceed 1/3 of the yield strength of the material in shear. Another constraint on the design is the angle of twist ϕ, which is not to exceed 4.4° measured from the motor to the gear C.

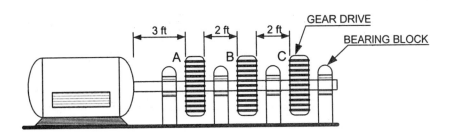

11.21 A torque of 20 ft-lb is applied to a structural member with an elliptical cross section that is fabricated from steel. The length of the structural member is 4.2 ft. The height h of the ellipse is 1.0 in. and its depth b is 0.35 in. Determine the maximum shear stress and the angle of twist for this structural member.

11.22 A torque of 25 N-m is applied to a steel structural member with an equilateral triangle for its cross section. The length of the structural member is 1.0 m. The equilateral triangle has a base b = 30 mm. Determine the maximum shear stress and the angle of twist for this structural member.

11.23 A torque of 30 ft-lb is applied to a steel structural member with a hexagon for its cross section. The length of the structural member is 6 ft. The inscribed diameter within the hexagon has a diameter of 0.75 in. Determine the maximum shear stress and the angle of twist for this structural member.

11.24 A torque of 800 N-m is applied to a steel structural member with a square cross section. The length of the structural member is 2.0 m. The side of the square section measures 40 mm. Determine the maximum shear stress and the angle of twist for this structural member.

11.25 Prove that the shear stress τ is zero at all four corners of a shaft with a square cross section that is subjected to a torque T.

11.26 Determine the eccentricity e locating the shear center for a beam fabricated from the channel section as shown in the figure to the right.

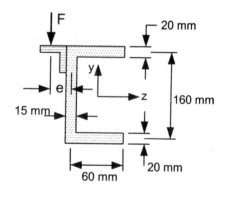

11.27 For the channel section, illustrated in the figure to the right, show that integrating the parabolic distribution of shear stresses over the area of the web yields a shear force V equal to:

$$V = F\left[\frac{2ct_1 + (3/5)ht_2}{2ct_1 + (2/3)ht_2}\right]$$

Explain why V is slightly less than F. Hint: the shear stress at the top of the web is given by:

$$\tau = F\left[\frac{ct_1 h}{t_2 I_z}\right]$$

CHAPTER 12 PROBLEMS

12.1 A three-dimensional stress state is defined by six Cartesian components of stress listed in the table below. Determine the state of strain, if the body is made from the material specified.

Prob. No.	σ_{xx}	σ_{yy}	σ_{zz}	τ_{xy}	τ_{yz}	τ_{zx}	Material
12.1a	15 ksi	21 ksi	−13.75 ksi	18 ksi	12.6 ksi	−10.75 ksi	Steel
12.1b	80 MPa	− 45MPa	92 MPa	35 MPa	− 72 MPa	46 MPa	Aluminum

12.2 A three-dimensional strain state is defined by six Cartesian components of strain listed in the table below. Determine the state of stress, if the body is made from the material specified.

Prob. No.	$\varepsilon_{xx} \times 10^{-6}$	$\varepsilon_{yy} \times 10^{-6}$	$\varepsilon_{zz} \times 10^{-6}$	$\gamma_{xy} \times 10^{-6}$	$\gamma_{yz} \times 10^{-6}$	$\gamma_{zx} \times 10^{-6}$	Material
12.2a	500	− 500	− 250	900	− 650	− 450	Steel
12.2b	1200	− 1450	800	1000	− 750	1100	Aluminum

12.3 The stresses on the external surface of a cylindrical pressure vessel, in the hoop and axial directions, are predicted to be σ_h = 35 ksi and σ_a = 17.5 ksi when the vessel is pressurized to 125 psi. If two strain gages are placed on the pressure vessel to measure the strains in the hoop and axial directions, predict the strains that are expected to occur at this pressure. The pressure vessel is fabricated from steel.

12.4 A plane state of stress at a point is represented by a small elemental area ΔA as shown in the figure to the right. The stresses are σ_{xx} = 15 ksi, σ_{yy} = 21 ksi and τ_{xy} = − 18 ksi. Determine the stresses $\sigma_{x'x'}$, $\sigma_{y'y'}$ and $\tau_{x'y'}$ on an element that is rotated (a) + 25° and (b) − 35° degrees (counterclockwise is positive) relative to the element shown in this figure.

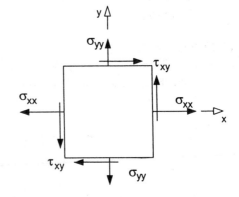

12.5 Consider a point, shown in the figure to the right, as an elemental area ΔA, with Cartesian components of stress σ_{xx} = 28 ksi, σ_{yy} = − 22 ksi and τ_{xy} = − 20 ksi. Determine the orientation of the principal planes and the stresses that act on these planes. Also determine the maximum shear stress, the corresponding normal stress, and the planes upon which they act.

12.6 Repeat Problem 12.5 for the following stress state: σ_{xx} = 100 MPa, σ_{yy} = − 55 MPa and τ_{xy} = 45 MPa. Determine the orientation of the principal planes and the stresses that act on these planes. Also determine the maximum shear stress, the corresponding normal stress, and the planes upon which they act.

12.7 Repeat Problem 12.4 using Mohr's circle for the solution.

12.8 Repeat Problem 12.5 using Mohr's circle for the solution.

12.9 Repeat Problem 12.6 using Mohr's circle for the solution.

12.10 Determine the pressure p required to cause yielding in a spherical pressure vessel if its radius is 800 mm and its wall thickness is 6.0 mm. The pressure vessel is fabricated from 80-20 brass. Use the maximum shear stress failure theory.

12.11 Determine the pressure p required to cause yielding in a cylindrical pressure vessel if its radius is 1.2 m and its wall thickness is 6.0 mm. The pressure vessel is fabricated from 304 stainless steel. Use the von Mises failure theory.

12.12 A pressure vessel, shown in the figure to the right, is employed as a water storage tank for a small municipality. Determine the height of the water, h, that can be maintained in the tank if the city's engineering director specifies a safety factor of 2.5. The tank is fabricated from 1020 HR steel with a 30.0 ft diameter, a height of 60.0 ft and a thickness of 1/8 in. The city's engineering director also indicates that the safety factor should be based on maximum normal stress theory for failure. Recall the density for water is 62.4 lb/ft^3.

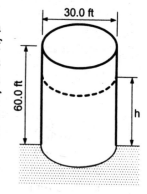

12.13 A cylindrical pressure vessel is fabricated by wrapping thin steel plate in a spiral about a mandrel with a diameter of 3.0 m and butt-welding the seam as illustrated in the figure to the

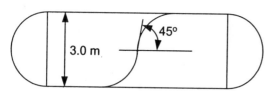

left. The weld seam forms an angle of 45° with the axis of the cylinder. Determine the normal stress σ and the shearing stress τ acting on the weld when the vessel is subjected to a pressure p = 700 kPa and is fabricated with 6.0 mm thick plate.

12.14 A 36 in. diameter pipeline is to be designed to transmit a variety of fluids. It is to operate at a maximum pressure of 500 psi. Determine the thickness of the steel plate used in its fabrication if the pipeline is designed with a safety factor of 3.3. The pipeline is fabricated from 1018 A steel. The design is based on von Mises theory for yielding.

12.15 A pair of orthogonal strain gages is mounted on a cylindrical pressure vessel with the orientation indicated in the figure to the right. The strains measured were ε_A = 538 × 10^{-6} and ε_B = 275 × 10^{-6}. The pressure vessel is fabricated from 1018 A steel plate that is 6 mm thick. Determine the hoop and axial stresses in the cylinder, the internal pressure

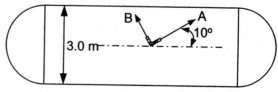

applied when the strains were measured and the factor of safety based on the von Mises theory for yielding.

12.16 A thin walled spherical water tank, presented in the figure to the right, is filled with water and vented to the atmosphere. Determine the meridional σ_m and tangential σ_t stresses anywhere along the equator of the sphere. Your result should be written as an equation in terms of the water density γ, the tank diameter D and its wall thickness t.

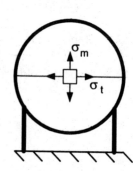

12.17 A state of plane strain exists at a specified point in a plane body where the Cartesian components of strain are $\varepsilon_{xx} = 800 \times 10^{-6}$, $\varepsilon_{yy} = 250 \times 10^{-6}$ and $\gamma_{xy} = -640 \times 10^{-6}$. Determine the state of strain on an element rotated through an angle $\theta = +25°$ (counterclockwise is positive) relative to the Oxy axes.

12.18 For the Cartesian components of strain $\varepsilon_{xx} = 950 \times 10^{-6}$, $\varepsilon_{yy} = -485 \times 10^{-6}$ and $\gamma_{xy} = 250 \times 10^{-6}$, determine the principal strains, the maximum shearing strains and the angles θ_p and θ_s. Also determine the normal strains on the planes of maximum shear strain.

12.19 Repeat Problem 12.17 using Mohr's strain circle for the solution.

12.20 Repeat Problem 12.18 using Mohr's strain circle for the solution.

12.21 A thin rubber membrane (dental dam) is stretched in such a manner that a uniform strain field is produced over the area of the membrane. The Cartesian components of strain are $\varepsilon_{xx} = 8,000 \times 10^{-6}$, $\varepsilon_{yy} = 2,500 \times 10^{-6}$ and $\gamma_{xy} = -6,000 \times 10^{-6}$. A rectangle is scribed on the membrane before it is deformed. Determine the orientation of the membrane relative to the Cartesian axes if its angles are to remain at 90° after stretching.

12.22 A thin aluminum plate 75 mm wide and 400 mm long is subjected to a state of stress that produces a uniform strain field with Cartesian components of $\varepsilon_{xx} = -720 \times 10^{-6}$, $\varepsilon_{yy} = 360 \times 10^{-6}$ and $\gamma_{xy} = 460 \times 10^{-6}$. Determine the change in length of the diagonals of the plate and determine the maximum shearing strain in the plate.

12.23 A two-element rectangular rosette is aligned with the principal stress directions. If the principal strain measurements are $\varepsilon_1 = 740 \times 10^{-6}$ and $\varepsilon_2 = -340 \times 10^{-6}$, determine the principal stresses. The structural element upon which the strain gages are mounted is fabricated from steel.

12.24 For the three-element rectangular rosette, derive Eqs. (12.64), (12.67) and (12.68).

12.25 A three-element rectangular rosette is mounted on an aluminum structure. The strain readings from each gage element are $\varepsilon_A = 1,800 \times 10^{-6}$, $\varepsilon_B = 600 \times 10^{-6}$ and $\varepsilon_C = -400 \times 10^{-6}$. Determine the principal strains, the principal stresses and the principal angle.

12.26 Prepare a spreadsheet that will automatically compute σ_1, σ_2, ε_1, ε_2 and θ_p from the strains ε_A, ε_B and ε_C obtained from measurements with a three-element rectangular rosette for three different materials (steel, aluminum and stainless steel).

12.27 Another three-element rosette in common usage, illustrated in the figure to the right, is the delta-rosette with angles $\theta_A = 0°$, $\theta_B = 120°$, and $\theta_C = 240°$. Determine the equation for the principal strains ε_1 and ε_2 in terms of ε_A, ε_B and ε_C for this rosette.

12.28 An axle of an all-terrain vehicle is subjected to the forces and torque shown in the figure to the left. If the axle diameter is 15 mm, determine the principal stresses and the maximum shearing stresses at point A and B on the surface of the axle. Note that point A lies along the z-axis and point B lies along the y-axis.

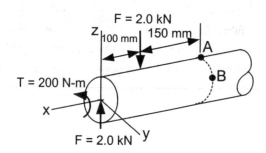

12.29 An advertising sign located along a city street is cantilevered from a pole as illustrated in the figure to the right. The pressure due to the wind impinging on the sign is uniformly distributed over its area and equal to 10 lb/ft² in the x direction. The pole is fabricated from a tube with an outside diameter of 8 in. and a wall thickness of 0.20 in. Determine the stresses at points A and B located at the base of the pole.

12.30 Repeat Problem 12.29; however, solve for the stresses at points C and D located at the base of the pole.

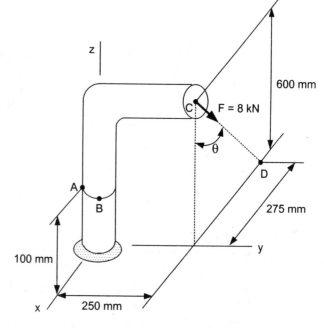

12.31 A machine component is fabricated from a tube that has been formed into a 90° bend as illustrated in the figure to the right. The outside diameter of the tube is 100 mm and its inside diameter is 88 mm. Determine the principal stresses and the maximum shear stresses at points A and B. Note that point A lies along the y-axis and point B lies along the x-axis.

12.32 A U shaped machine component, illustrated in the figure to the left, is subjected to opposing forces of 10 kN. If the component is fabricated from 1020 HR A steel, determine the safety factor based on yield strength using the maximum normal stress theory.

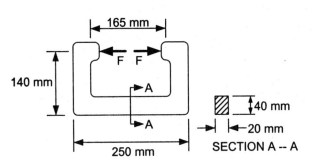

SECTION A -- A

12.33 Determine the maximum normal stress on the vertical section A -- A of the cantilever beam shown in the figure shown below.

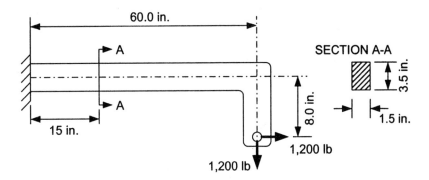

12.34 The large bracket, shown in the figure below, is loaded in its plane of symmetry with a force F = 4 kN inclined by an angle θ = 10°. Determine the principal stresses and the maximum shearing stresses at points A, B and C. Points A and C are in the web adjacent to the flange of the bracket.

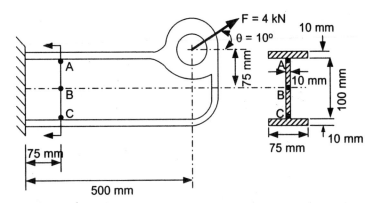

12.35 A thin walled cylindrical pressure vessel with an inner diameter 600 mm and a wall thickness of 10 mm, shown in the figure to the right, is subjected to an internal pressure of 1.0 MPa. It is also subjected to a torque of 50 kN-m and an axial force of 300 kN, which act through heavy plates attached to the ends of the cylinder. Determine the principal stresses and the maximum shear stresses at point A located on the outer surface of the cylinder.

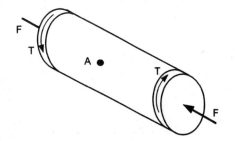

12.36 An angled strut 6 in. thick is subjected to a force F = 40 kip and a uniformly distributed load q = 10 kip/ft, as illustrated in the figure to the left. Determine the principal stresses at points A, B and C.

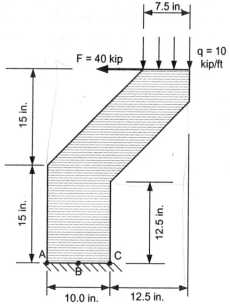

CHAPTER 13 PROBLEMS

13.1 For the cantilever beams shown in the figure below, write the equation for the elastic curves. Also determine the deflection and slope at the free end.

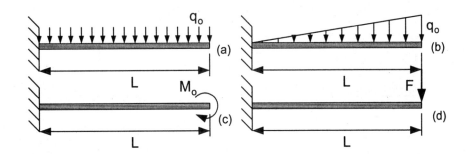

13.2 For the steel cantilever beam, shown in the figure to the right, determine the deflection and slope at the free end if the beam is fabricated from a W305 × 74 wide flange section.

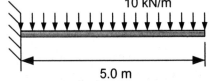

13.3 For the steel cantilever beam, shown in the figure to the left, determine the deflection and slope at the free end if the beam is fabricated from an American Standard section S254 × 38.

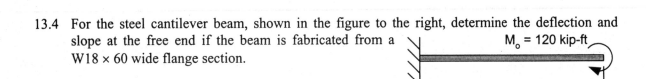

13.4 For the steel cantilever beam, shown in the figure to the right, determine the deflection and slope at the free end if the beam is fabricated from a W18 × 60 wide flange section.

13.5 For the steel cantilever beam, shown in the figure to the left, determine the deflection and slope at the free end if the beam is fabricated from an American Standard section S381 × 64.

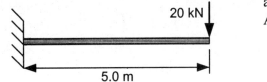

13.6   For the simply supported beams, shown in the figure below, write the equation for the elastic curves. Also determine the deflection at mid-span and the slope at both ends.

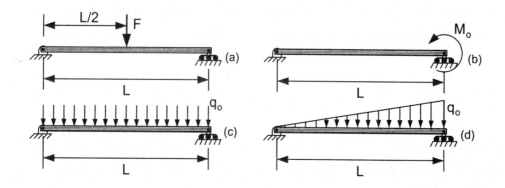

13.7   For the steel simply supported beam shown in the figure to the right, determine the mid-span deflection   and   slope   at   both   ends   if   the   beam   is fabricated from a S203 × 27 American Standard section.

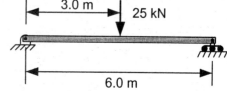

13.8   For the steel simply supported beam, shown in the figure to the left, determine the mid-span deflection and slope at both ends if the beam is fabricated from a W152 × 37 wide flange section.

13.9   For the steel simply supported beam, shown in the figure to the right, determine the mid-span deflection   and   slope   at   both   ends   if   the   beam   is fabricated from a W10 × 60 wide flange section.

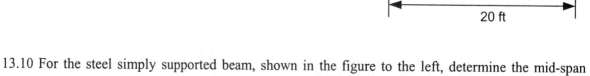

13.10 For the steel simply supported beam, shown in the figure to the left, determine the mid-span deflection and slope at both ends if the beam is fabricated   from   a   S254   ×   38   American Standard section.

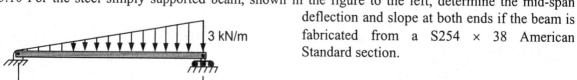

13.11 For the simply supported beam, shown in the figure to the right, determine the equation of the elastic curve and the slope at both ends.

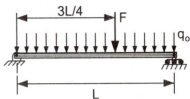

13.12 For the steel simply supported beam, shown in the figure to the left, determine the mid-span deflection and the slope at both ends if the beam is fabricated from a W10 × 60 wide flange section.

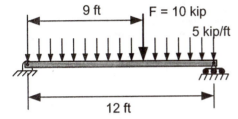

13.13 For the simply supported beam, shown in the figure to the right, determine the equation of the elastic curve and the slope at both ends.

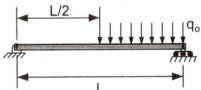

13.14 For the steel simply supported beam, shown in the figure to the left, determine the mid-span deflection and the slope at both ends if the beam is fabricated from a S18 × 70 American Standard section.

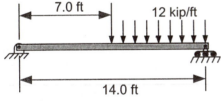

13.15 For the simply supported beam, shown in the figure to the right, determine the equation of the elastic curve and the slope at both ends.

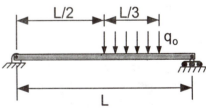

13.16 For the aluminum simply supported beam, shown in the figure to the left, determine the mid-span deflection and the slope at both ends if the beam is fabricated from a W457 × 89 wide flange section.

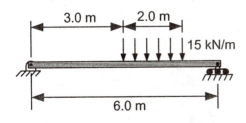

13.17 For the simply supported beam, shown in the figure to the right, determine the equation of the elastic curve and the slope at both ends.

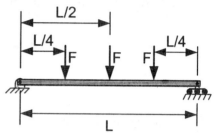

13.18 For the steel simply supported beam, shown in the figure to the left, determine the mid-span deflection and the slope at both ends if the beam is fabricated from a W203 × 46 wide flange section.

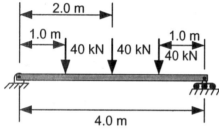

13.19 For the simply supported beam, shown in the figure to the right, determine the equation of the elastic curve and the slope at both ends.

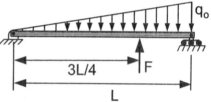

13.20 For the steel simply supported beam, shown in the figure to the left, determine the mid-span deflection and the slope at both ends if the beam is fabricated from a W533 × 92 wide flange section.

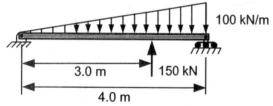

13.21 For the cantilever beam, shown in the figure to the right, write the equation for the elastic curve. Also determine the equations for deflection and slope at the free end.

13.22 For the steel cantilever beam, shown in the figure to the right, determine the deflection and the slope at the free end if the beam is fabricated from a S305 × 74 American Standard section.

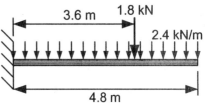

13.23 For the cantilever beam, shown in the figure to the left, write the equation for the elastic curve. Also determine the equations for the deflection and slope at the free end.

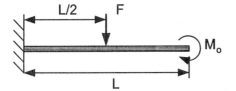

13.24 For the steel cantilever beam, shown in the figure to the right, determine the minimum weight wide flange section if the deflection at the free end is limited to 1.4 in.

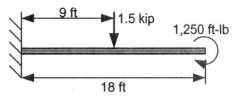

13.25 For the cantilever beam, shown in the figure to the left, write the equation for the elastic curve. Also determine the equations for the deflection and slope at the free end.

13.26 For the steel cantilever beam, shown in the figure to the right, determine the minimum weight American Standard section if the deflection at the free end is limited to 26 mm.

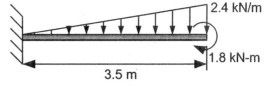

13.27 For the cantilever beam, shown in the figure to the left, write the equation for the elastic curve. Also determine the deflection and slope at the free end.

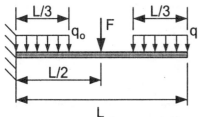

13.28 For the steel cantilever beam, shown in the figure to the right, determine the minimum weight structural tee section if the deflection at the free end is limited to 1.1 in.

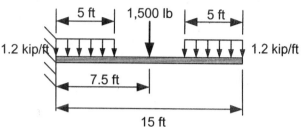

13.29 For the cantilever beam, shown in the figure to the left, write the equation for the elastic curve. Also determine the deflection and slope at the free end.

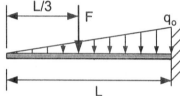

13.30 For the steel cantilever beam, shown in the figure to the right, determine the minimum weight wide flange section if the deflection at the free end is limited to 15 mm.

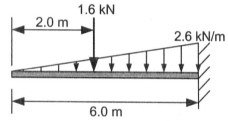

13.31 Using the method of superposition, for the simply supported beam shown in the figure to the left, determine the deflection at mid-span and the slope at the left end of the beam. The beam is steel with a W12 × 65 wide flange section.

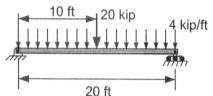

13.32 Using the method of superposition, for the simply supported beam shown in the figure to the right, determine the deflection at mid-span and the slope at the left end of the beam.

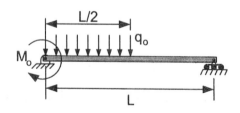

13.33 Using the method of superposition, for the simply supported beam shown in the figure to the left, determine the deflection at mid-span and the slope at the right end of the beam. The beam is steel with a W305 × 97 wide flange section.

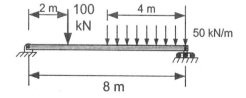

13.34 Using the method of superposition, for the simply supported beam shown in the figure to the right, determine the deflection at mid-span and the slope at the right end of the beam.

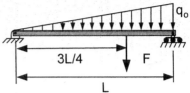

13.35 Using the method of superposition, for the cantilever beam shown in the figure to the left, determine the deflection and the slope at the free end of the beam.

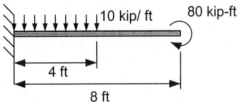

13.36 Using the method of superposition, for the cantilever beam shown in the figure to the right, determine the deflection and the slope at the free end of the beam. The beam is steel with a S457 × 81 American Standard section.

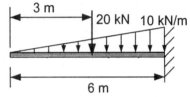

13.37 Using the method of superposition, for the cantilever beam shown in the figure to the left, determine the deflection and the slope at the free end of the beam. The beam is steel with a S20 × 66 American Standard section.

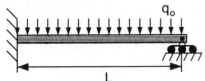

13.38 Using the method of superposition, for the cantilever beam shown in the figure to the right, determine the deflection and the slope at the free end of the beam.

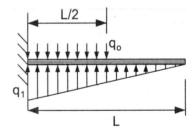

13.39 Determine the reactions at the prop and the wall on the cantilever beam shown in the figure to the left.

13.40 Determine the reactions at the prop and the wall on the cantilever beam shown in the figure to the right.

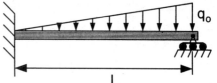

13.41 Determine the reactions at the prop and the wall on the cantilever beam shown in the figure to the left.

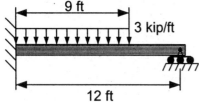

13.42 Determine the reactions at the prop and the wall on the cantilever beam shown in the figure to the right.

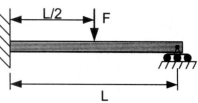

13.43 Determine the reactions at the prop and the wall on the cantilever beam shown in the figure to the left.

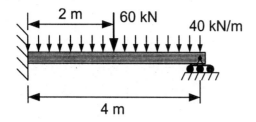

13.44 Determine the reactions at the prop and the wall on the cantilever beam shown in the figure to the right.

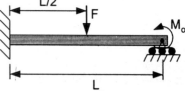

13.45 Determine the reactions at the supports of the continuous beam shown in the figure below.

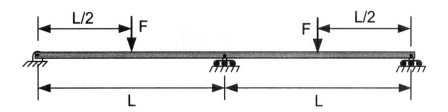

13.46 Determine the reactions at the supports of the continuous beam shown in the figure below.

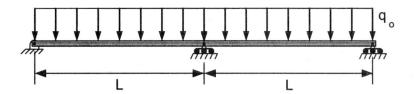

13.47 Determine the reactions at the supports of the continuous beam shown in the figure below.

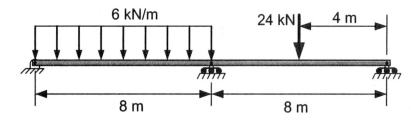

13.48 Determine the reactions at the supports of the continuous beam shown in the figure below.

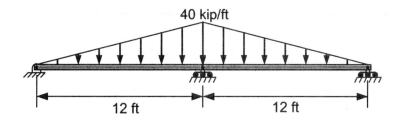

13.49 Determine the reactions at the supports of the continuous beam shown in the figure below.

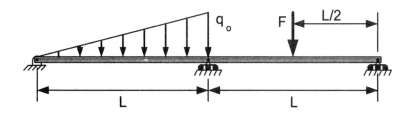

13.50 Determine the reactions at the supports of the built-in beam shown in the figure below.

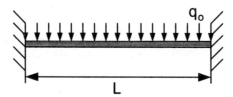

13.51 Determine the reactions at the supports of the built-in beam shown in the figure below.

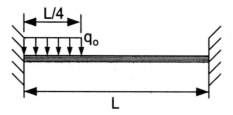

13.52 Determine the reactions at the supports of the built-in beam shown in the figure below.

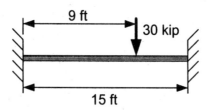

13.53 Determine the reactions at the supports of the built-in beam shown in the figure below.

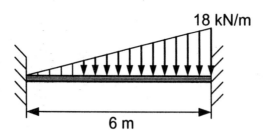

CHAPTER 14 PROBLEMS

14.1 Write an engineering brief explaining why understanding buckling is important in ensuring public safety.

14.2 An 18 ft long column is fabricated from an aluminum tube with an outside diameter of 8.0 in. and an inside diameter of 7.6 in. as shown in the figure to the right Determine the critical buckling force P_{CR} if the tube is pinned at both ends.

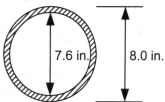

14.3 A 8 m long column is fabricated from a square steel tube as illustrated in the figure to the left. Determine the critical buckling force P_{CR} if the tube is pinned at both ends. Also determine the compressive stress in the tube at the critical load and comment on its magnitude.

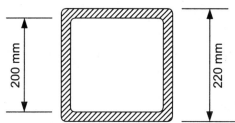

14.4 A 5 m long column is fabricated from an aluminum tube with the rectangular cross section defined in the figure to the left. The column is pinned at one end and built-in at the other. Determine the critical buckling force P_{CR}.

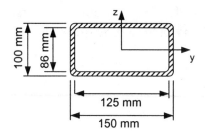

14.5 Design a round tubular column for a construction firm that is to be fabricated from aluminum. The column is 6.0 m long and is expected to exhibit a critical load of 300 kN. The ratio of its inside to outside diameter is 0.85. Both of its ends are supported by pins.

14.6 Design a column fabricated from an aluminum, square-tubular section for a construction firm. The column is 24 ft long and is expected to exhibit a critical load of 100 kip. The ratio of its inside to outside dimensions is 0.90. Both of its ends are supported by pins.

14.7 Design a column that is fabricated from a steel rectangular section. The column is 26 ft long and is expected to exhibit a critical load of 175 kip. The ratio of its inside to outside dimensions is 0.88 ($w_i/w_o = d_i/d_o = 0.88$). Both of its ends are built-in.

14.8 Select the most suitable wide flange section for the design of a steel column 7.1 m long. The column is expected to support a load of 500 kN and is built-in on both ends.

14.9 Select the most suitable American Standard section for the design of an aluminum column 16 ft long. The column is expected to support a load of 100 kip and is built-in on one end and pinned on the other.

14.10 Select the most suitable structural tee section for the design of a steel column 5.3 m long. The column is expected to support a load of 260 kN and is built-in on both ends.

14.11 A control rod in a vehicle is subjected to axial forces of ± 2,800 lb. If it is fabricated from stainless steel determine its diameter if the design incorporates a safety factor of 2.5. Consider both ends pinned.

14.12 A lever is used to provide an axial force to a 30 in. long vertical member as shown in the figure to the right. Determine the maximum force F that may be exerted if the vertical link is fabricated from an aluminum bar with a rectangular cross section. The dimensions of the rectangular bar are 15 mm by 40 mm. Note all of the joints are pinned.

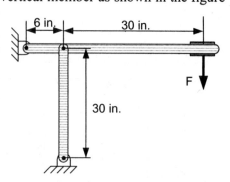

14.13 The truss shown in the figure to the left is fabricated from round, steel tubes. If the tubes all have the same outside diameter of 80 mm, determine the inside diameter of the compression members required to prevent buckling. Employ a safety factor of 4.0.

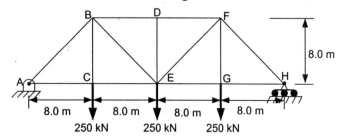

14.14 The truss, shown in the figure to the right, is fabricated from square, steel tubes. If the tubes all have the same outside dimension of s_o = 4.0 in., determine the inside dimension s_i of the compression members required to prevent buckling. Employ a safety factor of 2.5.

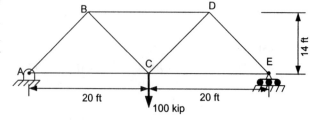

14.15 A force F acts on a simple truss containing a horizontal tie rod and an inclined strut, as shown in the figure to the right. The strut, fabricated from an aluminum alloy, is rectangular with dimensions w and d. Determine those dimensions if w = 2d and a safety factor of 2.5 is required.

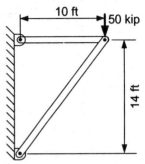

14.16 For the frame, shown in the figure to the left, determine the maximum force that can be supported before the inclined member buckles. The inclined member is fabricated from a rectangular steel bar with dimensions w and d where w/d = 1.5.

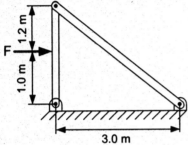

14.17 The truss, shown in the figure to the right, is fabricated from rectangular, steel tubes. If the tubes all have the same outside dimension of w_o = 150 mm in. and d_o = 75 mm determine the wall thickness of the compression members required to prevent buckling. Employ a safety factor of 4.0.

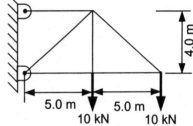

14.18 A wooden column, with a square cross section, is subjected to an eccentric load as shown in the figure to the left. The column is embedded in a concrete footer at one end and free at the other. Determine the load that can be applied to the column without causing it to fail by either buckling or breaking. The column is fabricated from Western white pine.

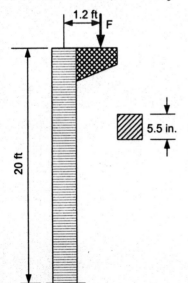

14.19 A column fabricated from a W305 × 74 wide flange section is loaded eccentrically as indicated in the figure to the right. Determine the force F required for the column to fail by either buckling or yielding. The wide flange section is fabricated from 1020 HR steel. The column is braced so that it does not buckle about its minimum inertia axis. Assume the top of the column is free.

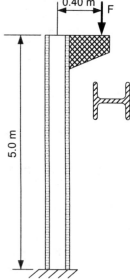

14.20 Determine the deflection y_{Max} at the top of the column, shown in the figure to the right, when the applied load F is equal to 80% of the failure load.

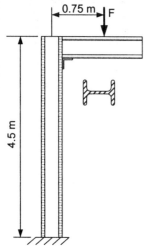

14.21 A column, fabricated from a S 203 × 34 American Standard section, is loaded eccentrically as indicated in the figure to the left. Determine the force F required for the column to fail by either buckling or yielding. The wide flange section is fabricated from 1020 HR steel. Assume both ends of the column are built-in, and that bracing prevents buckling about the minimum inertia axis.

14.22 A column is fabricated from a solid aluminum rod with a diameter of 60 mm as shown in the figure to the right. It is loaded with a force F with an eccentricity e = 30 mm. Determine the maximum force F that can be applied to the column before it fails by either buckling or yielding. The aluminum alloy employed is 2024-T-4.

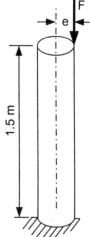

14.23 Develop a design curve similar to the one shown in Fig. 14.14 for steel columns with pinned/pinned ends and with a yield strength of 50 ksi. Also determine the limiting value of the slenderness ratio SR_c for this case.

14.24 Determine the safety factor as a function of (kL/r) that is employed using the design code for steel columns. Consider yield strengths of 35, 40, 45, and 50 ksi.

14.25 A 8.4 m long column is fabricated from a W 254 × 45 steel wide flange section with pinned/pinned ends. Use the curve resulting from the code equations that is shown in Fig. 14.14 to determine the allowable centric load applied to the column.

14.26 A 8.0 m long column is fabricated from a S 178 × 23 steel American Standard section with pinned/pinned ends. Use the curve resulting from the code equations that is shown in Fig. 14.14 to determine the allowable centric load applied to the column.

14.27 A 4.3 m long column is fabricated from aluminum (2014 T-6) in the form of a WT 178 × 36 structural tee section. Use the curve resulting from the code equations that is shown in Fig. 14.15 to determine the allowable centric load applied to the column. The ends are both built-in.

14.28 An aluminum (2014-T6) tube with a square cross section, shown in the figure to the right, is to be employed in a building lobby. The column length is 5.6 m and its ends are both built in. Determine the largest axial (centric) load that it can carry.

CHAPTER 15 PROBLEMS

15.1 An aluminum rod with a cross sectional area A = 400 mm² and length L = 1.0 m is subjected to an axial force of 32.5 kN as shown in the figure to the right. Determine the work **W** performed on the rod and the strain energy stored in the rod.

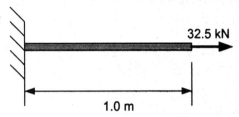

32.5 kN

1.0 m

15.2 A point in a machine component is subjected to a three-dimensional state of stress that is described in terms of three principal stresses σ_1 = 120 MPa, σ_2 = 85 MPa and σ_3 = – 99.4 MPa. Determine the strain energy density at this point if the machine component is fabricated from steel.

15.3 A point in a structural member is subjected to a three dimensional state of stress described by the six Cartesian stress components σ_{xx} = 60.2 MPa, σ_{yy} = 51.1 MPa, σ_{zz} = 34.2 MPa, τ_{xy} = – 25.3 MPa, τ_{yz} = 48.7 MPa and τ_{zx} =38.8 MPa. Determine the strain energy density at this point if the structural member is fabricated from brass.

15.4 A tie rod, 25 ft long with a 10 in. diameter is fabricated from steel. If the tie rod is subjected to an axial force of 1570 ton, determine the strain energy stored in the tie rod and the strain energy density.

15.5 For the simply supported beam shown in the figure below, determine the strain energy stored in it due to the normal stresses σ_{xx} produced by the uniformly distributed load q = 500 N/m. The beam is fabricated from steel.

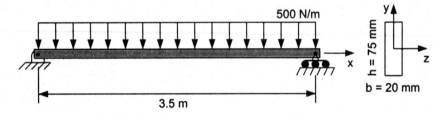

500 N/m

h = 75 mm

b = 20 mm

3.5 m

15.6 For the simply supported beam shown in the figure below, determine the expression for the strain energy due to the normal stress σ_{xx}. Assume that EI_z is a constant.

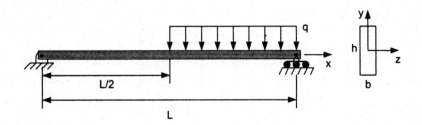

q

h

b

L/2

L

15.7 For the simply supported beam shown in the figure below, determine the expression for the strain energy due to the normal stress σ_{xx}. Assume that EI_z is a constant.

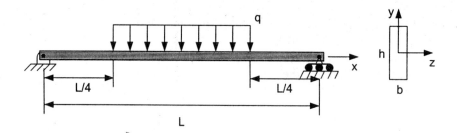

15.8 For the simply supported beam shown in the figure below, determine the expression for the strain energy due to the normal stress σ_{xx}. Assume that EI_z is a constant.

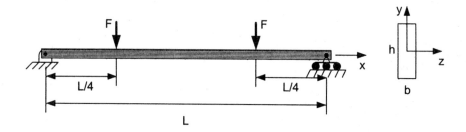

15.9 For the simply supported beam shown in the figure below, determine the expression for the strain energy due to the normal stress σ_{xx}. Assume that EI_z is a constant.

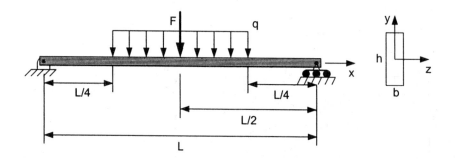

15.10 For the cantilever beam shown in the figure below, determine the expression for the strain energy due to the normal stress σ_{xx}. Assume that EI_z is a constant.

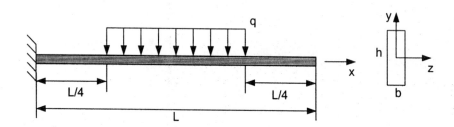

15.11 For the cantilever beam shown in the figure below, determine the expression for the strain energy due to the normal stress σ_{xx}. Assume that EI_z is a constant.

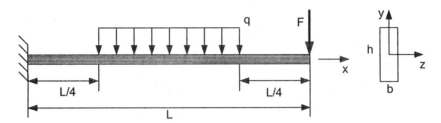

15.12 For the simply supported beam shown in the figure below, determine the relation for the strain energy stored due to the shear stress τ_{xy}. The beam has a rectangular cross section with dimensions h and b.

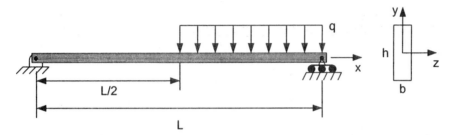

15.13 For the simply supported beam in shown in the figure below, determine the relation for the strain energy stored due to the shear stress τ_{xy}. The beam has a rectangular cross section with dimensions h and b.

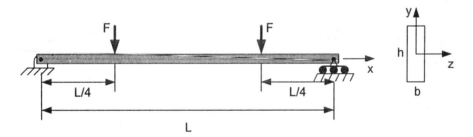

15.14 For the simply supported beam shown in the figure below, determine the relation for the strain energy stored due to the shear stress τ_{xy}. The beam has a rectangular cross section with dimensions h and b.

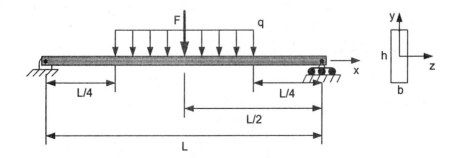

15.15 For the cantilever beam in shown in the figure below, determine the relation for the strain energy stored due to the shear stress τ_{xy}. The beam has a rectangular cross section with dimensions h and b.

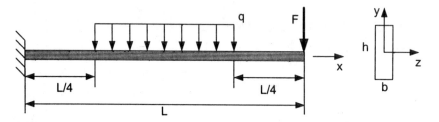

15.16 A 3.0 m long shaft with a 75 mm diameter, shown in the figure below, is subjected to a torque T = 450 N-m. Compute the strain energy stored in the shaft. The shaft is fabricated from brass.

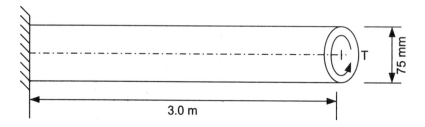

15.17 A stepped shaft, with lengths and diameters defined in the figure below, is subjected to a torque T = 110 ft-lb N-m. Compute the strain energy stored in the shaft. The shaft is fabricated from stainless steel.

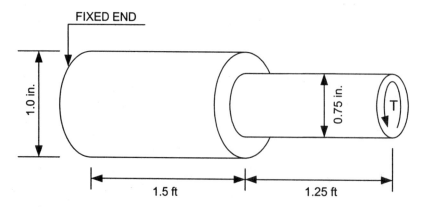

15.18 A sphere fired from an air gun strikes an aluminum rod with an axial impact as illustrated in the figure below. Determine the maximum force and the maximum stress produced if the sphere weighs 150 gm. The velocity of the sphere at the instant of impact is 200 m/s. The rod has a solid, circular cross section.

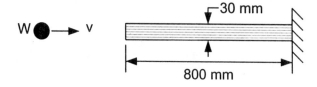

15.19 A short, steel cylinder of diameter d = 0.750 in. and length l = 2.25 in. is fired from an air gun and strikes a long, steel rod with an axial impact as shown in the figure below. Determine the maximum force and the maximum stress produced if the cylinder has a density of γ = 0.284 lb/in.^{3} The velocity of the cylinder at the instant of impact is 500 ft/s. The rod has a solid, circular cross section.

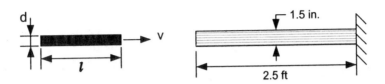

15.20 Consider a simply supported beam that is subjected to impact by a falling weight W as illustrated in the figure to the right. Derive the relation for the maximum force and the maximum bending stress in the beam. Assume that the deflection of the beam is small relative to the drop height H.

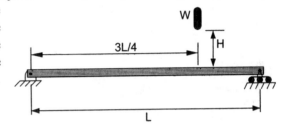

15.21 A weight W is positioned at the free end of a cantilever beam as shown in the figure to the left.

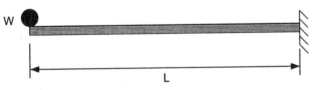

While the weight is in contact with the beam (H = 0), its gravitational force has not been released. The weight is suddenly released to produce an impact event, although the height of the drop is zero. Determine the equations for the maximum stress, the displacement of the load point and the maximum force developed during the impact.

15.22 For the simply supported steel beam shown in the figure to the right, determine the maximum force and the maximum stress when the weight impacts the beam. The beam is fabricated from a W 12 × 50 wide flange section.

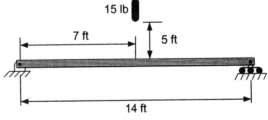

15.23 For the aluminum alloy cantilever beam shown in the figure to the left, determine the maximum force and maximum stress when the mass impacts the beam. The beam is fabricated from a S 127 × 22 American Standard section.

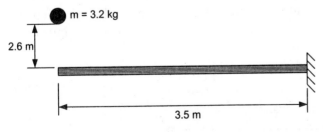

15.24 For the simply supported beam shown in the figure to the right, write the equation for the deflection at the quarter span (x = L/4) using Castigliano's theorem.

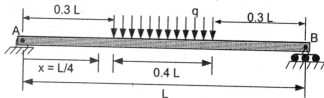

15.25 For the cantilever beam shown in the figure to the left, write the equation for the deflection at its free end using Castigliano's theorem.

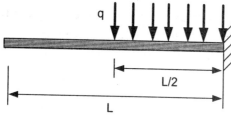

15.26 For the cantilever beam shown in the figure to the right, write the equation for the deflection at midspan and its free end using Castigliano's theorem.

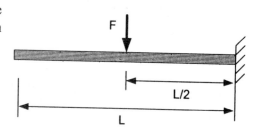

15.27 For the simply supported beam shown in the figure to the left, use Castigliano's theorem to write the equation for the deflection at the midpoint of the span.

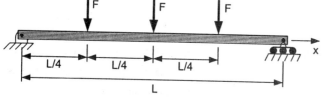

15.28 For the simply supported beam shown in the figure to the right, write the equation for the deflection at midspan and both quarter points using Castigliano's theorem.

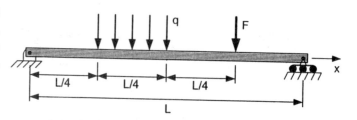

15.29 For the cantilever beam shown in the figure to the left, write the equation for the deflection at midspan and its free end using Castigliano's theorem.

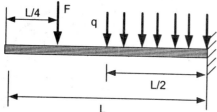

15.30 For the cantilever beam shown in the figure to the right, use Castigliano's theorem to write the equation for the reaction force at the prop.

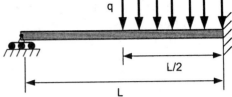

15.31 For the cantilever beam shown in the figure to the left, use Castigliano's theorem to write the equation for the reaction force at the prop.

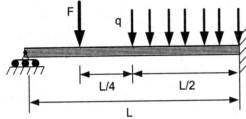

15.32 For the continuous beam shown in the figure to the right, use Castigliano's theorem to write the equation for the reaction force at the center support.

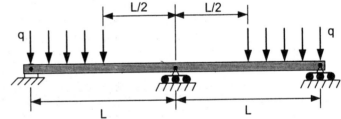

15.33 For the continuous beam, shown in the figure to the left, use Castigliano's theorem to write the equation for the reaction force at the center support.

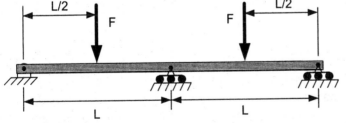

15.34 A cantilever beam is formed in the shape of a quarter circle as illustrated in the figure to the right. If the cantilever beam is loaded with a vertical force F at its free end, write the equation for the beam's deflection in the direction of the applied force. Assume that EI_z is a constant.

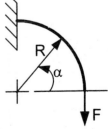

15.35 A cantilever beam is formed in the shape of a quarter circle as illustrated in the figure to the right. If the cantilever beam is loaded with a horizontal force F at its free end, write the equation for the beam's deflection in the direction of the applied force. Assume that EI_z is a constant.

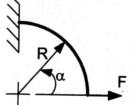

15.36 A cantilever beam is formed in the shape of a semicircle as illustrated in the figure to the right. If the cantilever beam is loaded with a horizontal force F at its free end, write the equation for the beam's deflection in the direction of the applied force. Assume that EI_z is a constant.

15.37 A cantilever beam is formed in the shape of a semicircle as illustrated in the figure to the right. If the cantilever beam is stretched by an amount δ in the horizontal direction, determine the force that must be applied to produce this deflection.

15.38 A shaft is fabricated by brazing together two sections of round tubing as shown in the figure below. The shaft is fixed at one end and a torque T is applied to it at its free end. Use Castigliano's theorem to write the equation for the angle of twist ϕ. Assume the thickness of the braze metal is negligible.

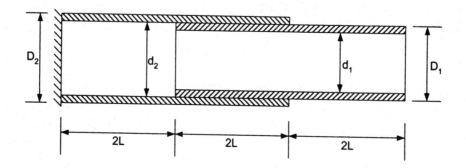

15.39 A shaft is fabricated by brazing together two sections of round tubing as shown in the figure below. The shaft is fixed at one end and a torque T is applied to it at its free end. Use Castigliano's theorem to write the equation for the angle of twist ϕ. Assume the thickness of the braze metal is negligible.

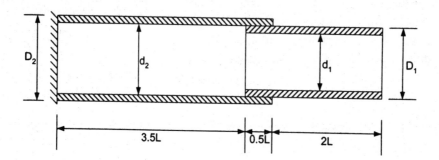

15.40 Write the equation for the displacement in the (a) vertical and (b) horizontal direction at joint B for the two-member truss, shown in the figure to the right, using Castigliano's method. Truss members AB and BC are identical of the same material, length L, and area A. The ratio of h/w is 4/3.

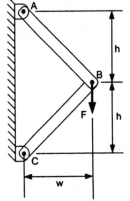

15.41 Determine the displacement in the (a) vertical and (b) horizontal direction at joint B for the two-member truss, shown in the figure to the right, using Castigliano's method. Truss members AB and BC are identical of the same material, length L, and area A. The ratio h/w = 1.00.

15.42 Write the equation for the displacement in the vertical direction at joint C for the seven-member truss, shown in the figure to the left, using Castigliano's method. All of the truss members are fabricated from the same material. The cross sectional area of members BC, AE, BD and DE is 2A; the area of AB is A; and the area of members BE and CD is 3A.

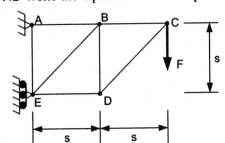

15.43 Write the equation for the displacement in the vertical direction at joint D for the five-member truss, illustrated in the figure to the right, using Castigliano's method. All of the truss members are fabricated from the same material with the same cross sectional area of A. The ratio h/s = 1.5.

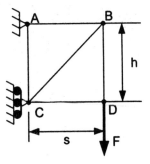

15.44 Write the equation for the displacement in the vertical direction at joint C for the seven-member truss, shown in the figure to the right, using Castigliano's method. All of the truss members are fabricated from the same material with the same cross sectional area of A. The ratio of h/s is 0.85.

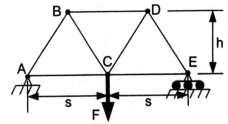

15.45 Write the equation for the displacement in the vertical direction at joint B for the seven-member truss, illustrated in the figure to the right using Castigliano's method. All of the truss members are fabricated from the same material with the same cross sectional area of A. The ratio of h/s is 0.85.

15.46 Write the equation for the displacement in the vertical direction at joint D for the five-member truss, shown in the figure to the right, using Castigliano's method. All of the truss members are fabricated from the same material with the same cross sectional area of A. The ratio of h/s is 1.20.

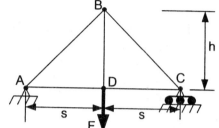

15.47 Write the equation for the displacement in the horizontal direction at joint C for the five-member truss, illustrated in the figure to the right, using Castigliano's method. All of the truss members are fabricated from the same material with the same cross sectional area of A. The ratio of h/s is 1.20.

CHAPTER 16 PROBLEMS

16.1 Your manager directs you to perform a stress analysis of a machine component containing a well defined crack, and to employ the von-Mises theory of yielding to predict when the structure will fail. How do you respond to this directive?

16.2 Determine the stress concentration at an elliptical hole of length 2a when a uniaxial stress σ_o is applied in a direction perpendicular to the length of the hole. Let b = a/50 and b = a/100. Where does σ_{max} occur?

16.3 Determine the stress concentration at an elliptical hole of length 2a when a uniaxial stress σ_o is applied in a direction parallel to the length of the hole. Let b = a/50 and b = a/100. Where does σ_{max} occur?

16.4 A crack 60 mm long develops in a very wide plate, which is to support a uniaxial stress of 70 MPa. Determine the stress intensity factor K_I. What assumption did you make in this determination?

16.5 If the crack in Problem 16.4 grows at a rate of 10 mm/month, determine the service life remaining before crack initiation if $K_{Ic} = 1265$ MPa√mm.

16.6 For a very wide plate subjected to a uniaxial stress of 100 MPa, prepare a graph showing the stress intensity factor as a function of crack length 2a.

16.7 Prepare a graph showing α as a function of (a/w) for a tension strip with a centrally located crack. Let (a/w) vary from 0.05 to 0.80.

16.8 A steel strap 2.0 mm thick and 20 mm wide with a central crack of 2.0 mm long is loaded to failure. Determine the critical load F_c if K_{Ic} for the strap material is 2780 MPa√mm.

16.9 Determine the failure stress σ_o applied to the strap of Problem 16.8. Compare this stress to the yield and tensile strengths of 4340 alloy steel. How will the strap fail?

16.10 Use a spreadsheet program to compute the multiplier α for SEC and DEC cracked tension strips. Evaluate α for 0 < (a/w) < 0.49 in increments of 0.01. Display the results in a listing similar to those shown in Table 16.2.

16.11 A steel tension bar, 6.0 mm thick and 20 mm wide with a single edge crack 3.0 mm long, is subjected to a uniaxial stress of 90 MPa. Determine the stress intensity factor K_I.

16.12 If K_{Ic} is 2,000 MPa√mm for the steel used to fabricate the tension bar in Problem 16.11, will the crack initiate?

16.13 Determine the critical crack length a_c for the tension bar in Problem 16.11 if $K_{Ic} = 2,000$ MPa√mm.

16.14 Determine the critical load F_c for the tension bar in Problem 16.11 if $K_{Ic} = 2,000$ MPa√mm.

16.15 A steel tension bar, 0.10 in. thick and 1.00 in. wide with double edge cracks each 0.15 in. long, is subjected to a uniaxial stress of 25 ksi. Determine the stress intensity factor K_I.

16.16 If K_{Ic} is 65 ksi√in. for the steel used to fabricate the tension bar in Problem 16.15, will the crack initiate?

16.17 Determine the critical crack length a_c for the tension bar in Problem 16.15 if K_{Ic} = 65 ksi√in.

16.18 Determine the critical load F_c for the tension bar in Problem 16.15 if K_{Ic} = 65 ksi√in.

16.19 Determine the normalized stress intensity factor $\dfrac{K_I bw^{3/2}}{Fs}$ as a function of (a/w) for a beam subjected to three-point bending. Let (a/w) vary from 0.05 to 0.50.

16.20 Determine the normalized stress intensity factor $\dfrac{K_I bw^2}{M\sqrt{a}}$ as a function of (a/w) for a beam subjected to pure bending. Let (a/w) vary from 0.05 to 0.50.

16.21 Find the critical load F_c that can be applied to a beam subjected to three-point bending as illustrated below. Note K_{Ic} = 1,500 MPa√mm.

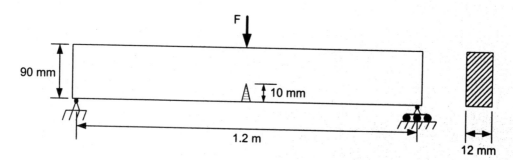

16.22 Find the critical moment $M_c = F_c d$ that can be applied to a beam in pure bending as illustrated in the figure below. Note K_{Ic} = 75 ksi √in.

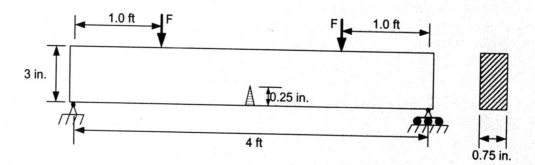

16.23 An inspector locates a 7.0 mm long crack adjacent to a 30 mm diameter circular hole in a tension panel, fabricated from aluminum 2124-T851. The tension panel is occasionally loaded with a uniaxial stress σ_o = 150 MPa. He recommends replacing the panel. You are to review his recommendation. Do you agree with his recommendation? Justify your answer.

16.24 Determine the critical crack length a_c for the geometry shown in the figure to the right. Note that $K_{Ic} = 55 ksi\sqrt{in}$.

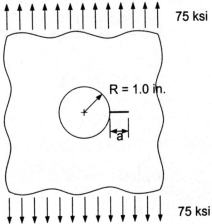

16.25 Determine the maximum size of an embedded shrinkage crack a_{Max} that can form and remain stable in an 18 Ni-steel casting if the crack is subjected to an applied stress of 800 MPa.

16.26 A surface flaw is introduced on an external corner of a 630-aluminum bronze component as shown in the figure to the right. Determine the applied stress σ_o the component can withstand before the crack (flaw) becomes unstable.

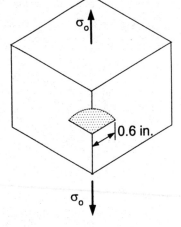

16.27 An inspector has located a shallow surface crack at the edge of hole in a tension panel as illustrated in the figure to the left. The crack dimensions are given in the figure. You are to establish the stress σ_o that will initiate the crack. Note the panel material is 70-titanium.

16.28 Give an example of:
  a. Opening mode loading.
  b. In-plane shear mode loading.
  c. Out-of-plane shear mode loading.
  d. Mixed mode (I and II) loading.

16.29 Determine r_p at the critical load for the beam shown in the figure below if $S_y = 700$ MPa. Compare r_p to the crack length and the ligament length. Comment on this comparison.

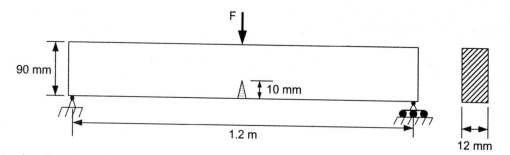

16.30 For the beam in shown below, which is the same as the one defined Problems 16.21 and 16.29, adjust the crack length to accommodate the effect of yielding if $S_y = 700$ MPa and then determine the new critical load F_c. Compare this result to that found in the original solution for Problem 16.21. Note that $K_{Ic} = 1,500$ MPa√mm.

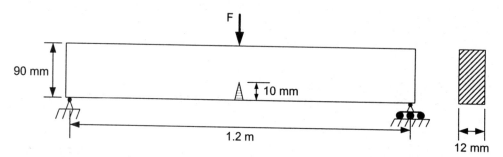

16.31 Consider a thin tension panel shown in the figure below. The panel is fabricated from a ductile material that is perfectly plastic when it yields at a stress of 120 ksi. Use the Dugdale model to determine the length of the plastic zone r_p.

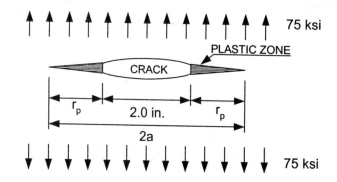

16.32 Determine the limit moment M_{Limit} that can be achieved prior to plastic rupture of a cracked beam subjected to pure bending. Use a spreadsheet to calculate M_{Limit}/S_y as a function of (a/w) as this ratio varies from 0 to 0.75.

16.33 Describe the effect of plate thickness on the fracture toughness of ductile metals.

16.34 Describe the effect of increasing the tensile strength of a metallic alloy on its fracture toughness.

16.35 Describe the effect of decreasing temperature on fracture toughness of typical structural steels.

16.36 Determine an approximate value for the fracture toughness of a structural steel if the Charpy V-notch test result for the absorbed energy is 25.0 J. The test temperature and the service temperature are both in the transition range.

16.37 A wide and thick steel plate contains an edge crack 0.65 in. long. The plate is subjected to alternating stresses between $\sigma_{Min} = 0$ ksi and $\sigma_{Max} = 22$ ksi. Experiments conducted with similar steel provided data for the material constants in Eq. (16.35) as $C = 5.8 \times 10^{-10}$ and $n = 4.0$ when ΔK is expressed in units of ksi$\sqrt{in}$. The plane strain fracture toughness for this steel is $K_{Ic} = 60$ ksi$\sqrt{in}$. Determine the number of cycles N_f for this crack to grow to a critical size that produces failure.

16.38 A wide and thick steel plate contains an edge crack 21 mm long. The plate is subjected to alternating stresses between $\sigma_{Min} = 0$ MPa and $\sigma_{Max} = 210$ MPa. Experiments conducted with similar steel provided data for the material constants in Eq. (16.35) as $C = 9.8 \times 10^{-11}$ and $n = 4.0$ when ΔK is expressed in units of MPa$\sqrt{mm}$. The plane strain fracture toughness for this steel is $K_{Ic} = 3,210$ MPa$\sqrt{mm}$. Determine the number of N_f cycles for this crack to grow to a critical size that produces failure.

16.39 Describe the three steps involved in fracture control plans.

16.40 Construct a graph showing the critical crack length a_c as a function of K_{Ic}/σ_o over the range from 0.01 to 0.15 $\sqrt{m}$. Assume $\alpha = 1$ for simplicity.

16.41 Write a fracture control plan for a large natural gas storage tank that is to be located adjacent to a densely populated metropolitan area.

16.42 Why is proof testing used to insure structural integrity?

APPENDIX C PROBLEMS

C.1 Determine the area of a right triangle if its base is 30 mm and its height is 70 mm.

C.2 Determine the area of an isosceles triangle if its base is 30 mm and its height is 70 mm.

C.3 Determine the area of a right triangle if its base is 4in. and its height is 7 in.

C.4 Determine the area of an isosceles triangle if its base is 5 in. and its height is 9.5 in.

C.5 Determine the area of a circle with a diameter of 14 mm.

C.6 Determine the area of a circle with a diameter of 3.2 in.

C.7 Determine the area of a circle with a diameter of 88 mm.

C.8 Determine the area of a circle with a diameter of 3.6 in.

C.9 Determine the area of an ellipse if the major and minor axes are defined by a = 2b = 16 mm.

C.10 Determine the area of an ellipse if the major and minor axes are defined by a = 3b = 6 in.

C.11 Determine the area of an ellipse if the major and minor axes are defined by a = 4b = 20 mm.

C.12 Determine the area of an ellipse if the major and minor axes are defined by a = (3.5)b = 7.0 in.

C.13 Draw a right triangle with a base of 30 mm and height of 70 mm and then locate and dimension its centroid.

C.14 Draw an isosceles triangle with a base of 30 mm and height of 70 mm and then locate and dimension its centroid.

C.15 Draw a right triangle with a base of 4in. and height of 7 in. and then locate and dimension its centroid.

C.16 Draw an isosceles triangle with a base of 5 in. and height of 9.5 in. and then locate and dimension its centroid.

C.17 Verify the location of the centroid for the semicircular area defined in the figure to the right.

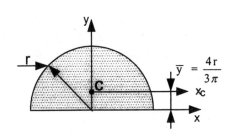

C.18 Verify the location of the centroid for the semi-elliptical area defined in the figure to the right.

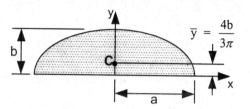

C.19 Verify the location of the centroid for the parabolic area defined in the figure to the left.

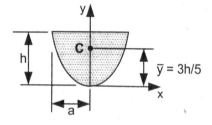

$\bar{y} = 3h/5$

C.20 Verify the location of the centroid for the circular sector defined in the figure to the right.

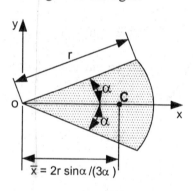

$\bar{x} = 2r \sin\alpha /(3\alpha)$

C.21 Verify the location of the centroid for the general spandrel defined in the figure to the left.

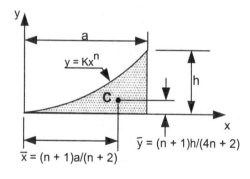

$y = Kx^n$

$\bar{y} = (n + 1)h/(4n + 2)$

$\bar{x} = (n + 1)a/(n + 2)$

C.22 For the unusual plate shown in the figure to the right, derive an equation locating the centroid relative to point A. The elliptical hole is centered in the rectangular portion of the plate.

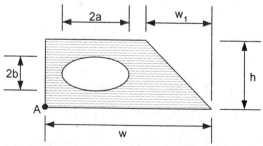

C.23 Determine the location of the centroid relative to point A for the plate defined in the figure to the right. All dimensions are given in mm, and the elliptical hole is centered in the rectangular portion of the plate.

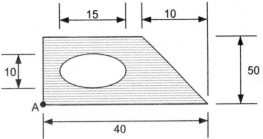

C.24 Divide the plate, illustrated in the figure to the right, into three areas (an ellipse, rectangle and triangle). Determine the moment of inertia for each area about its centroidal axis.

C.25 Verify that the moment of inertia of an ellipse about its centroidal axis is given by:

$$I_x = \pi ab^3 /4 \qquad \text{and} \qquad I_y = \pi a^3 b/4$$

C.26 Determine the moment of inertia for the total area of the plate, shown in the figure to the left, about its centroidal axis.

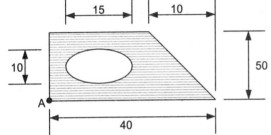

C.27 For the U shaped section, shown in the figure to the right, determine the location of the centroid $\bar{y}$ relative to an axis along its base. All dimensions are in inches.

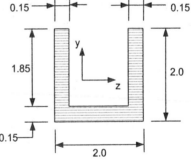

C.28 Determine the location of the centroid $\bar{y}$, relative to an axis along its base, for the U shaped section shown in the figure to the left. All dimensions are in inches.

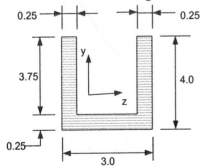

C.29 Determine the moment of inertia I_z about the centroid of the U shaped member defined in (a) Problem C.27 and (b) Problem C.28.

C.30 Verify that the polar moment of inertia of a quarter circle area with a radius R is $J_o = \pi R^4/8$.

C.31 For the uncommon shape presented in the figure to the right, determine the centroid and the moment of inertia relative to the centroidal axis. The rectangular hole is centered in the elliptical area.

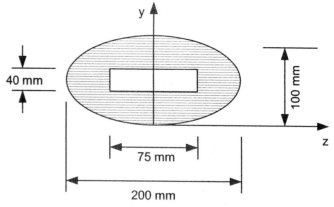

C.32 For the uncommon shape presented in the figure to the right, determine the moment of inertia relative to the z-axis.

C.33 For the uncommon shape presented in the figure to the right, determine the moment of inertia relative to the y-axis.

C.34 For the irregular cross sectional area, described in the figure to the left, locate the centroid relative to both the x and y axes. The dimensions are given in mm.

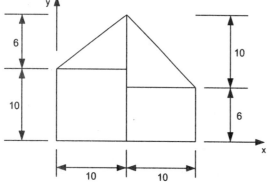

ANSWERS

Chapter 1

1.2	103.0 N on Earth
	39.02 N on Mars
1.3	$W_s = 11.43$ N
1.14	$p = 152.8$ psi
1.16	$d = 2.985$ in.
1.21(a)	$F = 10,680$ N
1.21(f)	$F = 422.6$ kN
1.22(a)	$F = 502.2$ lb
1.22(c)	$F = 2.968$ kip
1.27(a)	$\sigma = 82.74$ MPa
1.27(d)	$\sigma = 913.6$ MPa
1.28(a)	$\sigma = 145.0$ ksi
1.28(d)	$\sigma = 203.0$ psi

Chapter 2

2.16(a)	$S_v = 8,660$ N, $\theta_{Sv} = 90°$
2.16(b)	$D_v = 13,230$ N, $\theta_{Dv} = -40.9°$
2.16(c)	$D_v = 13,230$ N, $\theta_{Dv} = 139.1°$
2.17(a)	$S_v = 913.9$ N, $\theta_{Sv} = 70.04°$
2.17(b)	$D_v = 3,093$ N, $\theta_{Dv} = -46.93°$
2.17(c)	$D_v = 3,093$ N, $\theta_{Dv} = 133.1°$
2.18	$F_x = 15,105$ lb
	$F_{y'} = -3,202$ lb
2.19	$F_{AB} = 1,949$ N; $F_{CB} = 1,647$ N
2.21	$S_v = 15.61$ kN; $\theta = 209.7°$
2.22	$F = 8.640$ kN; $\theta = 46.67°$
2.25	$F_1 = 7,072$ lb; $\theta = 49.0°$

2.26	$F_1 = 10,120$ lb; $\theta = 55.87°$
2.33	$M_O = 438.8$ N-m (CCW)
2.34	$M_O = 86.6$ N-m (CCW)
2.37(a)	$\bar{r} = (7\bar{i} + 4\bar{j} + 6\bar{k})$ in.
	$r = 10.05$ in.
	$\alpha = 45.85°$, $\beta = 66.55°$, $\gamma = 53.34°$
2.37(b)	$\bar{r} = (110\bar{i} + 65\bar{j} + 70\bar{k})$ mm
	$r = 145.7$ mm
	$\alpha = 40.98°$, $\beta = 63.50°$, $\gamma = 61.29°$
2.39(b)	$F_x = 388.2$ lb, $F_y = 964.2$ lb
	$F_z = 1082$ lb
	$\bar{F} = (388.2\bar{i} + 964.2\bar{j} + 1082\bar{k})$ lb
2.39(d)	$F_x = 16.07$ kip, $F_y = -19.15$ kip
	$F_z = 0$
	$\bar{F} = (16.07\bar{i} - 19.15\bar{j})$ kip
2.40	$\bar{F} = \bar{F}_1 + \bar{F}_2 = (26.25\bar{i} + 37.23\bar{j} - 15.9\bar{k})$ kN
2.41(a)	$\theta = 87.35°$
2.41(b)	$\theta = 18.28°$
2.43(a)	$\bar{M}_O = (-6.4\bar{i} + 14.95\bar{j} - 7.85\bar{k})$ kN-m
2.43(b)	$\bar{M}_O = (-14.5\bar{i} - 13.28\bar{j} + 3.5\bar{k})$ ft-kip
2.45	$x_B = -58.62$ ft; $y_B = -3.307$ ft
2.47	$\bar{F} = (5.057\bar{i} + 2.528\bar{j} - 20.23\bar{k})$ kN
	See equation above for F_x, F_y, F_z.
	$\bar{M}_O = (-80.91\bar{i} + 20.23\bar{j} - 17.70\bar{k})$ kN-m
	See equation above for M_x, M_y, M_z.
	$M_O = 85.26$ kN-m; $\theta = 28.15°$
	$\bar{u}_{M_0} = (-0.949\bar{i} + 0.237\bar{j} - 0.208\bar{k})$
	$F_{OQ} = 18.52$ kN(Compression)
2.48	$\bar{F} = (741.8\bar{i} + 3,338\bar{j} - 6,677\bar{k})$ lb
	See equation above for F_x, F_y, F_z.

2.48 Continued:

$\overline{M}_O = (-73.44\overline{i} + 33.38\overline{j} + 8.530\overline{k})$ ft - kip

See equation above for M_x M_y, M_z.

$M_O = 81.12$ ft-kip

$\overline{u}_{Mo} = (-0.9053\overline{i} + 0.4115\overline{j} + 0.1052\overline{k})$

$\theta = 36.09°$;

$F_{OQ} = 6.061$ kip (Compression)

Chapter 3

3.6 $F_{CB} = 220.9$ lb; $F_{CA} = 585.3$ lb

3.7 $F_{AC} = 11.97$ kN; $F_{BC} = 9.807$ kN

3.9 $T_{AC} = 1,189$ N; $T_{BC} = 1,158$ N

3.10 $F = 1.879$ ton
        Pulley system is not effective.

3.11 $F = 4.358$ kN
        Pulley system is effective.

3.15 $F = W(1 - (r_2/r_1))/2$

3.16 $F_{AB} = 421.7$ N; $F_{BD} = 730.4$ N;
        $F_{CG} = 1461$ N; $F_{BC} = F_{CE} = 843.4$ N

3.18 $T_{AB} = 186.0$ N; $T_{BC} = 173.9$ N
        $T_{CD} = 212.0$ N; $m_C = 7.413$ kg

3.19 $W_{max} = 6,200$ lb > 140 lb
        Stuntman will remain dry.

3.21 $\delta_T = 3.851$ in.; $k_{eff} = 6.076$ lb/in.

3.22 $\delta_T = 48.57$ mm; $k_{eff} = 1.441$ N/mm

3.23 $\delta_T = 24.36$ mm; $k_{eff} = 47.00$ N/mm

3.26 $A_x = 0$; $A_y = 10$ kN; $B_y = 30$ kN

3.27 $A_x = 0$; $A_y = 1,406$ lb; $B_y = 2,344$ lb

3.28 $A_x = 0$; $A_y = 26.16$ kN; $B_y = 53.16$ kN

3.29 $A_x = 0$; $A_y = 11$ kip;
        $M_{RA} = 121$ kip-ft (CCW)

3.30 $A_x = 0$; $A_y = 37.33$ kN;
        $M_{RA} = 161.8$ kN-m (CCW)

3.31 $A_x = 625.8$ lb; $A_y = 491.8$ lb
        $B_y = 559.0$ lb

3.33 $A_x = 0$; $A_y = B_y = 40$ kN

3.36 $W_{max} = 1.892$ ton

3.37 $R_x = 0$; $R_y = 38.25$ kN
        $M_z = 860.6$ kN-m

3.38 $A_x = -13.5$ kip; $B_x = 7.5$ kip;
        $A_y = B_y = 4.5$ kip

3.39 $P_A = 36$ kN; $P_B = 18$ kN

3.40 $V = F/4$; $M = FL/8$

3.42 $V = q_0L/6$; $M = q_0L^2/9$

3.44 $V = -F$; $M = -(3/4)$ FL

3.46 $V = -4$ kip; $M = 76.8$ kip-ft

3.48 $P = 250$ lb, $V = 0$; $M_z = -375$ in-lb

Chapter 4

4.1 $\sigma = 19.10$ ksi; $\varepsilon = 0.0006367$
        $L_f = 22.014$ ft

4.4 $\sigma_{max} = 1.104$ MPa, $\sigma_{12} = 764.8$ kPa
        $\sigma_{23} = 320.7$ kPa

4.7 $P_{design} = 11.77$ kN

4.9 $\sigma = 1744$ psi

4.13   4340 Hr — Gage # 4-Os
        52100 A — Gage # 4-Os

4.18 $\sigma = 22$ MPa; $\delta = 0.1913$ mm

4.19 $\sigma_{design} = 16.88$ ksi; $A = 1.185$ in.^{2}

4.22 $\tau = 21.33$ ksi

4.25 $P_{max} = 1.651$ kip

4.27 $\sigma_\phi = (6.365)\sin^2 \phi$ MPa
      $\tau_\phi = (6.365)\sin \phi \cos \phi$ MPa

4.28 $\sigma = 90.0$ MPa; $\theta = 161.6°$
      $F = 15.75$ kN

4.32 $\delta = 2.235$ mm

4.35 $\sigma_1 = 30.30$ ksi; $\sigma_2 = -1.538$ ksi
      $\delta = 0.01841$ in.

4.36 $\sigma_{nom} = 20$ ksi; $\sigma_{max} = 46.6$ ksi

4.37 $\sigma_{nom} = 25$ MPa; $\sigma_{max} = 43.75$ MPa

4.41 $P_{max} = 343.5$ kN

4.42 $t_{min} = 0.320$ in.

4.43 $\sigma_a = -174$ MPa; $\sigma_f = -20$ kPa;
      $\delta = 4.833$ mm (shorter)

4.44 $\sigma_s = 33.03$ ksi; $\sigma_b = -8.256$ ksi
      $\delta = 0.006192$ in. (shorter)

4.45 $\sigma_s = -7.761$ ksi; $\sigma_c = -1.164$ ksi;
      $\delta = 0.02483$ in (shorter)

4.46 $\sigma_a = 19.04$ MPa; $\sigma_b = 761.7$ kPa;
      $\delta_a = 0.1983$ mm (longer);
      $\delta_b = 0.001731$ mm (longer)

4.47 $\sigma_c = -20.49$ MPa; $\sigma_s = -512.1$ MPa
      $\delta_c = 0.4647$ mm (longer)
      $\delta_s = 0.4647$ mm (shorter)

4.48 $\sigma_s = 1.784$ ksi; $\sigma_a = -467$ psi

Chapter 5

5.5 $S_y = 46.1$ ksi; $S_u = 62.2$ ksi
     $S_y (0.2\%) = 56.7$ ksi

5.6 $S_y = 36$ ksi; $S_u = 44$ ksi
     $S_y (0.2\%) = 38.5$ ksi

5.7 $\%e = 32.8\%$; $\%A = 31.38\%$

5.9 $d = 0.4997$ in.

5.12 $V = V_0 [1 + (1 - 2\nu)\varepsilon_a] = V_0$
      when $\nu = \frac{1}{2}$ and higher order strain terms
      are neglected.

5.13 $\varepsilon_y = 0.001267$

5.14 $\varepsilon_a = 0.003153$; $\varepsilon_t = -0.001009$

5.15 $\varepsilon_x = \varepsilon_y = 0.0004807$

5.18 $\sigma_a = 52.8$ MPa; $\sigma_h = 105.6$ MPa

5.19 $\sigma = 57.55$ ksi; $\sigma_T = 107.4$ ksi

5.23 $N = \infty$ infinite life in fatigue

5.26 $\sigma_{max} = 255$ MPa

5.33 $E = 30,000$ ksi; $G = 11,490$ ksi
      $\nu = 0.3054$

5.34 $E = 9.888 \times 10^6$ psi, $\sigma_u = 59.08$ ksi
      $\sigma_f = 48.89$ ksi; $\varepsilon_f = 0.2100$
      $\sigma_y (0.2\%) = 40.21$ ksi
      $\%e = 21.00\%$; $\%A = 31.77\%$

5.35 $E = 199.6$ GPa, $\sigma_u = 300.0$ MPa
      $\sigma_f = 150.0$ MPa; $\varepsilon_f = 0.1864$
      $\sigma_y (0.2\%) = 199.6$ MPa
      $\%e = 18.64\%$; $\%A = 49.78\%$

Chapter 6

6.8 $P_{AC} = 26.25$ kip (T); $P_{BC} = 52.5$ kip (T)
     $P_{CD} = 36.34$ kip (C); $P_{CE} = 42.5$ kip (T)

6.10 $P_{AC} = 67.48$ kip (T); $P_{BC} = 135.0$ kip (T)
      $P_{CD} = 67.08$ kip (C); $P_{CE} = 97.48$ kip (T)

6.12 $A_{AB} = 6.261$ in.^{2}; $A_{AC} = 2.8$ in.^{2}
      $A_{BD} = 2.8$ in.^{2}; $A_{BC} = 5.6$ in.^{2}

6.13 $A_{AB} = 1,933$ mm^2; $A_{AF} = 1,694$ mm^2;
      $A_{BC} = 1,607$ mm^2; $A_{BF} = 285.7$ mm^2

6.17 $P_{DE} = 0$; $P_{DF} = P_{DB} = 30$ kip (C)

6.20 $P_{DC} = 100$ kN (T); $P_{DE} = 131.7$ kN (T)
      $P_{DF} = 257.1$ kN (C)

6.23 $P_{AC} = 200(s/h)$;
$P_{AD} = -200[(s/h)^2 + 1]^{1/2}$

6.25 $P_{AC} = 3.907$ kN (T); $P_{CD} = 0.2504$ kN (C)
$P_{CE} = 4.206$ kN (T); $P_{DF} = 4.206$ kN (C)

6.26 $P_{CD} = 41.08$ kN (C); $P_{CJ} = 11.27$ kN (T)
$P_{KJ} = 34.53$ kN (T)

6.30 $\sigma_{BC} = 10.4$ ksi (T), $SF_{BC} = 4.038$

6.35 $A_{AB} = 4{,}340$ mm^2; $A_{BD} = 4{,}340$ mm^2
$A_{DE} = 5{,}482$ mm^2; $A_{EF} = 2{,}856$ mm^2
$A_{CF} = 2{,}284$ mm^2

Chapter 7

7.4 $P_{EB} = -8.944$ kips; $d_{EB} = 1.033$ in.
$P_{EC} = -4.472$ kips; $d_{EC} = 0.7305$ in.
$P_{ED} = 15.56$ kips; $d_{ED} = 1.363$ in.

7.13 $MOS_{AB} = 165.1\%$
$MOS_{AP} = MOS_{AQ} = 181.1\%$

7.14 $d_{AB} = 8.148$ mm; $d_{AC} = 7.374$ mm
$d_{AD} = 13.98$ mm

7.17   Gage # 2; steel wire

7.18 $P_{AD} = 165.9$ lb; $P_{BD} = 437.6$ lb
$P_{CD} = 441.1$ lb

7.19   Gage # 000s steel wire

7.20 $W = 1007$ lb; $A_x = -419.6$ lb
$A_y = 2{,}518$ lb; $A_z = -881.2$ lb
$P_{BC} = 1{,}730$ lb; $P_{DE} = 1{,}841$ lb

7.21 $F = 104.2$ kN; $A_x = 41.7$ kN
$A_y = 0$; $A_z = 250$ kN
$B_x = -145.9$ kN; $B_z = 250$ kN

Chapter 8

8.4 $F = 36.84$ N

8.5 $G_y = 2{,}591$ N; MA = 8.637

8.12 $P_C = 30$ ton; MA 6

8.13 $F_C = 165$ lb; $\boldsymbol{W} = 8.4$ in-lb
$\Delta d_C = 0.0509$ in.

8.14 $A_x = -4.5$ kip; $A_y = 7.3$ kip
$M_{RA} = 64.35$ kip-ft; $C_y = -1.443$ kip
$C_x = -12.75$ kip; $P_{BD} = -15.46$ kip

8.15 $B_y = 3.84$ kN; $C_x = 0$; $C_y = -9.6$ kN;
$E_x = 0$; $E_y = 0.96$ kN; $P_{AF} = 0$

8.17 $P_{AC} = 1{,}178$ N; $P_{BC} = -834.4$ N
$P_{CE} = 627.9$ N; $D = 699.3$ N

8.21 $C_x = 2.5$ kip; $C_y = 14.59$ kip; $A_x = 0$;
$A_y = 34$ kip; $M_{RA} = 50$ kip-ft (CCW);
$T = 13.35$ kip; $F = 15.61$ kip

Chapter 9

9.5(a) $\mu_{Ave} = 0.4156$; $\mu_{Range} = 0.0515$

9.5(b) $\mu_{Ave} = 0.4530$; $\mu_{Range} = 0.0656$

9.7 $F = 1{,}845$ N

9.9 $(F_{fW})_{max} = 57.95$ lb $< F_{fW} = 64.48$ lb
Worker slips and cannot move crate.

9.11 $(F_1)_{max} = 1094$ lb

9.12 $F_{fA} = 26.17$ lb < 57.93 lb $= (F_{fA})_{max}$
The ladder is stable.

9.13 $F = 240$ N; The crate begins to tip.

9.14 $(F_{fA})_{max} = 156.4$ N < 210 N $= F_{fA}$
Cylinder slips and remains in corner.

9.30 $F_{fB} = 21.25$ lb < 50 lb $= (F_{fB})_{max}$
$x_B = 0.4808$ ft < 1.5 ft
$F_{fA} = 21.25$ lb < 38.25 lb $= (F_{fA})_{max}$
Both blocks are in equilibrium.

9.31 $(W_1)_{max} = 1.667$ kN; $(W_1)_{min} = 0.3$ kN

Chapter 10

10.3 $\varepsilon_{max} = 0.002660$; $\sigma_{max} = 191.5$ MPa

10.4 $M_{max} = 480$ kN-m

10.5 $\sigma_A = -2.17$ ksi; $\sigma_B = -0.7232$ ksi
       $\sigma_C = 1.446$ ksi; $\sigma_D = 1.808$ ksi

10.6 $M_{max} = 117.5$ kip-in (+ moment)
       $M_{max} = 249.6$ kip-in (– moment)

10.7 $I_z = 394.6 \times 10^6$ mm^4
       $Z = 1,902 \times 10^3$ mm^3

10.9 $I_z = 2,113$ in.^{4}; $I_y = 115.2$ in.^{4}

10.10 $I_z = 3,013$ in.^{4}; $I_y = 176.0$ in.^{4}

10.11 $M_{Modified}/M_{Original} = 2.106$

10.14 $V_1(x) = 6.444$ kip; $M_1(x) = 6.444x$ kip-ft
       $V_2(x) = 1.444$ kip;
       $M_2(x) = [1.444x + 30]$ kip-ft
       $V_3(x) = -6.556$ kip;
       $M_3(x) = [-6.556x + 118]$ kip-ft

10.15 $V(x) = -10$ kN; $M(x) = -10x$ kN-m

10.16 $V(x) = [3.15 - x]$ kN
       $M(x) = [3.15x - x^2/2]$ kN-m

10.17 $V_1(x) = [5.192 - 0.07x]$ kip
       $M_1(x) = [5.192x - 0.035x^2]$ kip-ft
       $V_2(x) = [-1.808 - 0.07x]$ kip
       $M_2(x) = [35 - 1.808x - 0.035x^2]$ kip-ft

10.18 $V_1(x) = 1.6$ kN
       $M_1(x) = [1.2 + 1.6x]$ kN-m
       $V_2(x) = [3.4 - 1.2x]$ kN
       $M_2(x) = [-0.15 + 3.4x - 0.6x^2]$ kN-m
       $V_3(x) = -2$ kN
       $M_3(x) = [12 - 2x]$ kN-m

10.19 $Z_{min} = 984.1 \times 10^3$ mm^3
       W 406 × 60 for minimum weight beam

10.20 $Z_{min} = 750 \times 10^3$ mm^3
       S381 × 64 for minimum weight beam

10.21 $Z_{min} = 52.22 \times 10^3$ mm^3
       WT 203 × 19 for minimum weight beam

10.22 $\tau_{max} = 1.5$ ksi

10.23 $\tau_{max} = 16.98$ MPa

10.24 $\tau_a = 1.2$ MPa; $\tau_{max} = 1.4$ MPa

10.26 $s = 1.767$ in.

10.28 $q_y = 5,734$ N/m

10.30 $h_f = 245.6$ mm; $t_m = 0.5080$ mm
       for h/b ratio = 3/1

10.31 $(M_z)_{max} = 692.7$ kip-in.

10.32 $(M_y)_{max} = 120$ kip-in.

10.35 MOS = 4.167%

10.36 $\sigma_c = 20.82$ MPa (C); $\sigma_s = 156.2$ MPa (T)

10.44 $\sigma_{nom} = 85.5$ ksi; $\sigma_{max} = 179.6$ ksi

10.46 $\sigma_{nom} = 71.29$ MPa; $\sigma_{max} = 114.1$ MPa

Chapter 11

11.2 $\gamma_{max} = 312.2$ $\mu\varepsilon$

11.4 $\tau_{max} = 17.39$ ksi

11.6 $d_0 = 14.26$ mm

11.10 $\tau_{nt} = 4.676$ ksi; $\sigma_n = 4.515$ ksi

11.13 $\phi = 3.464°$

11.15 $\tau_{max} = 847.5$ MPa in DE
       $\phi_{AE} = 32.83°$; $\phi_{AC} = -3.871°$

11.17 $\tau_{nom} = 15.65$ ksi; $\tau_{max} = 18.47$ ksi

11.18 $T_{max} = 215.5$ N-m

11.19 $d = 20$ mm; $D = 27$ mm
       $r = 3$ mm for 4340 HR steel

11.20   d = 0.875 in.; D = 1.75 in.
        r = 0.175 in.

11.21 τ_{max} = 9.978 ksi; ϕ = 8.036°

11.24 ϕ = 3.213°; τ_{max} = 60.07 MPa

Chapter 12

12.1(a) ε_x = 427.5 $\mu\varepsilon$; γ_{xy} = 1,565 $\mu\varepsilon$
        ε_y = 687.5 $\mu\varepsilon$; γ_{yz} = 1,096 $\mu\varepsilon$
        ε_z = − 818.3 $\mu\varepsilon$; γ_{zx} = − 934.8 $\mu\varepsilon$

12.2(a) σ_x = 7.212 ksi; τ_{xy} = 10.35 ksi
        σ_y = − 15.87 ksi; τ_{yz} = − 7.475 ksi
        σ_z = − 10.10 ksi; τ_{zx} = − 5.175 ksi

12.4(a) $\sigma_{x'}$ = 2.283 ksi; $\sigma_{y'}$ = 33.72 ksi;
        $\tau_{x'y'}$ = − 9.272 ksi

12.5 σ_1 = 35.02 ksi; σ_2 = − 29.02 ksi
        θ_1 = − 19.33°; θ_2 = 70.67°
        τ_{max} = 32.02 ksi; σ_n = 3 ksi
        θ_s = 25.67°

12.6 σ_1 = 112.1 MPa; σ_2 = − 67.12 MPa;
        θ_1 = 15.07°; θ_2 = − 74.93°;
        τ_{max} = 89.62 MPa; σ_n = 22.5 MPa;
        θ_s = − 29.93°

12.10 p_{max} = 750 kPa

12.12   h = 26.92 ft

12.13 σ_w = 131.3 MPa; τ_w = − 43.75 MPa

12.14 t_{min} = 0.8037 in.

12.17 $\varepsilon_{x'}$ = 456.5 $\mu\varepsilon$; $\varepsilon_{y'}$ = 593.4 $\mu\varepsilon$
        $\gamma_{x'y'}$ = − 832.7 $\mu\varepsilon$

12.20 ε_1 = 960.8 $\mu\varepsilon$; ε_2 = − 495.8 $\mu\varepsilon$;
        θ_1 = 4.941°; θ_2 = − 85.06°;
        γ_{max} = 1,457 $\mu\varepsilon$; ε_n = 232.5 $\mu\varepsilon$;
        θ_s = 49.94°

12.23 σ_1 = 21.03 ksi; σ_2 = − 3.890 ksi

12.25 ε_1 = 1,805 $\mu\varepsilon$; ε_2 = − 404.5 $\mu\varepsilon$;
        σ_1 = 19.41 ksi; σ_2 = 2.004 ksi;
        θ_1 = − 2.597°

12.29 σ_A = 41.80 ksi; τ_A = 5.045 ksi
        σ_B = 0; τ_B = − 5.502 ksi

12.30 σ_C = − 41.80 ksi; τ_C = − 5.045 ksi
        σ_D = 0; τ_D = 4.588 ksi

12.31   At point A:
        σ_1 = 43.24 MPa; σ_2 = − 1.085 MPa
        τ_{max} = 22.16 MPa
        At point B:
        σ_1 = 41.06 MPa; σ_2 = − 2.737 MPa
        τ_{max} = 21.90 MPa

12.32 SF_y = 1.222

12.33 σ_{max} = 14.73 ksi

12.34   At point A:
        σ_1 = 1.741 MPa; σ_2 = − 0.1628 MPa
        τ_{max} = 0.9519 MPa
        At point B:
        σ_1 = 1.838 MPa; σ_2 = − 0.2618 MPa
        τ_{max} = 1.050 MPa
        At point C:
        σ_1 = 1.737 MPa; σ_2 = − 0.1631 MPa
        τ_{max} = 0.9501 MPa

12.35 σ_1 = 32.29 MPa; σ_2 = − 2.944 MPa
        τ_{max} = 17.62 MPa

12.36   At point A:
        σ_1 = 0; σ_2 = − 11.24 ksi
        At point B:
        σ_1 = 0.9493 ksi; σ_2 = − 1.053 ksi
        At point C:
        σ_1 = 11.04 ksi; σ_2 = 0

Chapter 13

13.1(b) $y(x) = \dfrac{q_0}{120LEI}\left(-20L^3x^2 + 10L^2x^3 - x^5\right)$

$y_B = -\dfrac{11q_0L^4}{120EI}$

$y'_B = -\dfrac{q_0L^3}{8EI}$

13.5 $y_B = -21.64$ mm, $y'_B = -0.3720°$

13.6 $y(x) = \dfrac{M_0}{6LEI}\left(x^3 - L^2x\right)$

$y_{mid} = -\dfrac{M_0L^2}{16EI}$ $\qquad y'_A = -\dfrac{M_0L}{6EI}$

$y'_B = \dfrac{M_0L}{3EI}$

13.14 $y_C = -0.1867$ in.; $y'_A = -0.1782°$
$y'_B = -0.2292°$

13.15

$y(x) = \dfrac{q_0}{1944EI}\left(36Lx^3 - 81\left\langle x - \dfrac{L}{2}\right\rangle^4 + 81\left\langle x - \dfrac{5L}{6}\right\rangle^4 - 31L^3x\right)$

$y'_A = -\dfrac{31q_0L^3}{1944EI}$ $\qquad y'_B = \dfrac{19q_0L^3}{972EI}$

13.20 $y_C = -0.2543$ mm; $y'_A = -0.01534°$
$y'_B = 0.005482°$

13.23

$y(x) = \dfrac{1}{EI}\left(\dfrac{F}{6}x^3 - \left(\dfrac{FL}{4} + \dfrac{M_0}{2}\right)x^2 - \dfrac{F}{6}\left\langle x - \dfrac{L}{2}\right\rangle^3\right)$

$y_B = -\dfrac{L^2}{48EI}\left(5FL + 24M_0\right)$

$y'_B = -\dfrac{L}{8EI}\left(FL + 8M_0\right)$

13.28 $I_{min} = 312.4$ in.^{4}; WT 15×54
for minimum weight beam

13.32

$y_C = -\dfrac{L^2}{768EI}\left(5q_0L^2 + 48M_0\right)$

$y'_A = -\dfrac{L}{384EI}\left(9q_0L^2 + 128M_0\right)$

13.34

$y_C = -\dfrac{L^3}{768EI}\left(5q_0L + 11F\right)$

$y'_B = \dfrac{L^2}{5760EI}\left(128q_0L + 315F\right)$

13.35

$y_B = -\dfrac{7FL^3}{16EI}$ $\qquad y'_B = -\dfrac{5FL^2}{8EI}$

13.38

$y_B = \dfrac{L^4}{3840EI}\left(128q_1 - 70q_0\right)$

$y'_B = \dfrac{L^3}{48EI}\left(2q_1 - q_0\right)$

13.39 $A_y = (5/8)q_0L$; $B_y = (3/8)q_0L$
$M_{RA} = -(1/8)q_0L^2$

13.44 $A_y = (11/16)F + (3/2)(M_0/L)$
$B_y = (5/16)F - (3/2)(M_0/L)$
$M_{RA} = -(3/16)FL - (1/2)M_0$

13.49 $A_y = (2q_0L/15) - (3/32)F$
$B_y = (2q_0L/5) + (11/16)F$
$C_y = -(q_0L/30) + (13/32)F$

13.51 $A_y = (121q_0L/512)$
$B_y = (7q_0L/512)$
$M_{RA} = -(67q_0L^2/3072)$
$M_{RB} = -(13q_0L^2/3072)$

Chapter 14

14.2 $P_{CR} = 82.06$ kip

14.4 $P_{CR} = 340.7$ kN

14.8 $(I_y)_{min} = 3.084 \times 10^6$ mm^4; W127 × 24
for minimum weight column

14.12 $P_{max} = 513.8$ lb

14.15 $d = 4.422$ in.; $w = 8.844$ in.

14.17 $I_{Required} = 2.570 \times 10^6$ mm^4
$t_{min} = 6.558$ mm

14.18 $F_{max} = 4.900$ kip (Buckling)

14.21 $F_{max} = 95.05$ kN (Yielding)

Appendix C

C.17 $\bar{x} = 0$ $\bar{y} = \dfrac{4r}{3\pi}$

C.18 $\bar{x} = 0$ $\bar{y} = \dfrac{4b}{3\pi}$

C.23 $\bar{x} = 17.81$ mm $\bar{y} = 23.72$ mm

C.25 $I_x = (\pi/4)ab^3$; $I_y = (\pi/4)ba^3$

C.27 $\bar{y} = 0.7241$ in

C.28 $\bar{y} = 1.554$ in.

C.29(a) $I_z = 0.3535$ in^4

C.29(b) $I_z = 4.343$ in.^{4}

C.34 $\bar{x} = 9.445$ mm $\bar{y} = 6.278$ mm